Acknowledgments

I would never have been able to finish this book without the guidance of ALLAH, , help from friends, and support from my family.

I would like to express my deepest gratitude to Prof. Dr. Kamel H. Rahouma for his guidance and precious comments that helped me.

I would like to express my deepest gratitude to my company TE for giving the chance to me to get a great experience in the optical transmission network, helping and encouraging me to use its management lab and the data sets of its infrastructure of the optical network from different vendors.

Finally, I would like to express my gratitude and appreciation to my family, colleagues, and friends for their continuous patience, love, and support. My sincere gratitude and thanks go to my teachers who have been the guiding light of my life. I express my heartfelt gratitude and love to my parents, my wife, and my sons for encouraging and motivating me during difficult times.

Contents

Contents 1

Chapter 1 3

Introduction 3

1.1 Introduction 3

1.2 Existing challenges 6

1.3 Objectives and Contributions of This book 7

1.4 Book organization: 8

Chapter 2 10

The Advanced Technologies in the Optical Network 10

2.1 Introduction 10

2.2 Optical Transport Network 10

2.3 Software-Defined Network Fundamentals 15

2.4 The Machine Learning Technology in the Optical Network 18

Chapter 3 22

Review of the Major Challenges in the backhaul of the 4G Network 22

3.1. Introduction 22

3.2 The Challenges of 4G Network 23

    3.1.3. Integration with IP Multimedia Subsystem (IMS) 31

3.2. Security in backhaul section of LTE Network 33

3.3. Capacity in the backhaul section of LTE Network 40

3.4. Energy consumption in LTE 4G Network 43

3.5. Timing synchronization of LTE Network 47

3.3 Conclusion 52

Chapter 4 53

Intelligent Security of the 4G Traffic over the Optical Transport Network 53

4.1 Introduction 53

4.2 The Structure of the OTN Frames in the Optical Network 55

4.3.1 Intrusion Detection and Response Model with machine learning 58
4.3.2 Intrusion Detection in the physical layer of the optical network.. 59
4.3.3 The Proposed Security Layer 61
4.4 Analysis of the Proposed Security Model 66
5.1 Introduction 69
5.2. The operations of the optical network 70
5.2.1 Monitor the total Energy consumption over all-optical networks 71
5.2.2 The time of the fault localization 71
5.2.3 The Optical performance monitoring 72
5.2.4 Configuration management 72
5.3 The Proposed Solution for Smart Operations in the Optical Network 72
5.4 Conclusion 80
6.1 Introduction 81
6.2 Resilience Challenges and Motivations 82
5.3 Proposed Resilience Model in the Long Distance OTN 86
5.3.1 Developing the Physical Infrastructure of the Optical Network .. 86
5.3.2 Building the SDN Orchestration 88
5.3.3 The experience of resilience network behaviours 89
5.4 Network scenarios and analysis 91
5.6 Conclusion 94
7.1 Testing the Security Model 95
7.1.1 Test the intrusion detection and response Model 95
7.1.2 Testing the proposed security layer in the OTN 96
7.2 Results of the Power Consumption model 98
7.3 Results of the Performance model 100
7.4 Results of the Resilience model 101
8.1 Conclusion 103
References 104
List of Figures 109
List of Tables 112
Acronyms 113

Chapter 1

Introduction

1.1 Introduction

Today and shortly, the demands to transfer a tremendous mass of data among the various applications in many fields inside the country become an exclusive responsibility particularly with the significant developments of the clouding policies in the different forms such the data centers and the implementations of many smart applications of the 5G in the various areas of the information technologies. The current design of the core communication network consists of 3 independent zones, the 1'st zone is layer 1 and layer 2 which recognized as the physical layer of the core optical network such Optical Transport Network (OTN), the 2'nd band is layer 3 which viewed as IP core network, and the last zone is the access and the application layers such 4G network. The challenge in the current communication model is no associations among these 3 zones of the communication network to optimize the capacities and the needed resources which are required to carry the data among any of them. Now the optical transport network (OTN) represents the critical point of carrying a tremendous amount of data for the mobile network such 4G between the different sites according to the requirements of the users of the mobile operators [1]. The construction of the optical transport network (OTN) in our case is a case study of a long distances optical network, the study consists of multiple domains of transmission equipment's from different vendors, and joined with each other's by thousands of kilometers from the optical fiber cables, at the same time 1 or 2 separate domains cover each region inside the country from the long-distance OTN with the same or another vendor types. The problem in this traditional structure of the long-distance optical transport network is the average quality of the service's resilience and the challenge to reuse the free bandwidths on a particular optical area domain. This technique in the structure of the optical network converted it to several isolated domains of the optical network with chaotic merging between the legacy layers such as the synchronize digital hierarchy (SDH) and the new generation layers as the OTN equipment.

Furthermore, the routing among the various OTN regions demands more hand-operated and hardware orders to achieve it, which is not decent, while any failure in the core OTN with the significant obligations to restore the affected services. The new generations of the communication markets for the next 3 years will include the clouding thoughts in the several fields of the communication markets, and will increase in the implementations of much smart applicability following the employing of the 5G technologies around the world, all of these enormous progress in the communication markets in the near future will need extraordinary requirements in the infrastructure of the core optical network, which depends mainly on the performance, the latency, the available bandwidth, the dynamically of the recovery of the service from specific paths to others, the switching rate and other parameters in the OTN and the security of the traffic over the optical network [2].

The structure of the communication network between the different operators, which is built-in isolated zones between accesses network, IP core network, and the optical system makes the cost the operations is very high, which affects the quality of the service in the 4G network.

By suggesting the associations between the software-defined network (SDN), the machine learning technology (ML), and the conventional network management systems (NMS's) of the OTN the transformation of the existing determined OTN to be more dynamic and automated, this will be done by employing only one centralized controller (CC) to manage the OTN in the multiple vendors' fields, several layers and in the domains of the various regions. The idea of the optical network cloud is proposed between the various network areas to grant the traffic repairs between the different vendors' equipment in the IP and the OTN domains.

Running in the operational tasks of the optical network 4G by the conventional systems is very costly and immediately affects the net gains of any telecom operator. The Optimization of the operational expenses optical network needs to employ the operational tasks in intelligent methods by utilizing the proper machine learning techniques. The most critical factors which affect the cost of the operations in an optical network are how they can control and regulate the energy losses, decrease the time of the fault localization, observe the quality of the transmission links periodically and lastly determining the most desirable routes for the formulations of the new circuits to decrease the consumption of the limited network resources. The book suggests, for the first-time universal platform which manages the essential operation factors in the optical network to the automation, and this produced by the synergies between four various advanced models to automate the achievements of the operational tasks in the system. The Forms of the operational functions are performed by the different techniques of the Artificial Neural Network (ANN) to overcome the human interventions in the achievements of these tasks.

Although the client data signals move through several layers of the optical transmission network to terminate its targets, the bulk of the telecom operators employ conventional encryption algorithms to secure those signals at the application layers only. Nevertheless; any interested intruder can reach the physical layer of the transmission network from any location of the optical fiber connections and he can divide the optical cables by wiretapping it, at that case the attacker will have a live representation of the optical signals, and by trying little attempts to solve out the conventional patterns of the OTN system, the attacker can read the first layer of the customer data signals [3]. A newly suggested model for adding a new security layer in the OTN frame structure is introduced to secure the customer's data over the optical system with the help of the artificial intelligence (AI) prediction feature and the software-defined network (SDN) of a centralized controller. The AI and the SDN model will reply to any intrusion discoveries in the optical system by enabling the new security layer within the mapping steps of the OTN frame in the Network Element (NE), and it will be employed to do the required encryption algorithm for the customers' signals according to operator decision.

The existing challenges in the backhaul of the 4G networks and the Optical Transport Networks (OTN) affect the performance, the flexibility, and the cost of building the next generation of the communication network. At the same time, the solutions to these challenges will lead to the evolution of the 5G mobile networks, which are characterized by an increasing number of wireless devices, service complexity, and the high demands to access more mobile services. The book discusses the impact of the implementations of the new technologies such software-defined network and machine learning in the core optical network on the flexible functionality of the access networks and how the centralized controller in the heterogeneous networks could improve the capacity and the resilience of the 4G services. There are many challenges of the LTE network in the backhaul section and the core OTN, such as the integration between the different heterogamous optical networks, synchronization of the network, the security of the customer data over the core optical network, and the available capacity of the backhaul LTE network. One of these changes is the security of the mobile customers over the backbone optical network such as OTN. The proposed security model in this book is implemented on the basis of protecting the vital client signals only over the optical layers by passing these signals into extra layer called security layer, this done by adding a new card in the Network Element (NE) to perform this job and by using the concept of the software-defined network (SDN) of the centralized controller with the help of the machine learning technology to perform the automatic detection of any intrusions over the optical layers. Another challenge in building the 4G network is the high cost of the leased lines over the core optical network, which is used to connect the 4G sites between the different regions in the country. In this book, for the first time an Intelligent Universal Platform (IUF) is proposed to manage and optimize the operational tasks in the optical network. Use cases in two situations As-Is and To-Be are studied, the cases are about the energy consumptions, the fault locations, the circuit creations and finally the variation in the signal to noise ratio. The results show that by using the machine learning (ML) in our platform the time of the fault location, the efforts to create one circuit, the number of the complaints, and the response time to the customer complaints are enhanced by significant percentages, this indicates that the machine learning techniques will play a significant role in monitoring, detecting, localizing the faults, and finally optimizing the resources of optical core network, all without human interventions in the near future.

There are many other difficulties in the OTN, such the expensive cost of the multilayer services planning, the quality of the services, and the quality of the resilience, all of these difficulties must be recovered first to cope with the changes in the new generations of the access communication networks. The needs to overcome many of these difficulties become vital nowadays, and depend on many factors in the OTN, such the status of the optical cables, the flexibility, the responsive, and the availability of OTN assets to direct customer control. In this book for a new proposed model is introduced by reorganizing the OTN resources to fit the needs of the new generations of the communications market, the model consolidates two promising technologies with each other which are the Software-Defined Network (SDN) and the Machine Learning (ML) to overcome the previous challenges and to reconstruct the control of the traditional OTN to be more smart, virtualized and automated.

For the first time, the optical cloud concepts are introduced in the OTN by slicing and virtualizing the various domains with its vendors in the heterogeneous optical network.

The results of the proposed OTN model according to the practical case study in the long-distance heterogamous OTN show that:

The dependence on the single vendor is nearly neglected with applying the concept of the clouding and slicing in the heterogeneous OTN, the pay for the end-users bandwidths has become possible and the time to provide the bandwidth on demand has become short , the meshing between the heterogeneous optical network became available and the resilience for diamond services improved, the available bandwidth of the optical core network in the long-distance network is optimized, the revenue from some OTN domains which have free bandwidths more than 50 % is increased, the switching time was enhanced, and the latency was reduced for the selected services which are optimized from the centralized controller.

1.2 Existing challenges

In our book we considered the optical core network is part of the 4G network, while the 4G network faces many problems in its back-hauling and front-hauling, as it needs to reserve leased lines over the optical transport network to connect the different sites of the 4G over all the country, the limitations of the leased lines affect directly the performance of the mobile system especially the 4G traffic, and without overcoming these limitations it will be complicated to satisfy the requirements of the 5G implementation. As the critical impact of the availability and the performance of the optical core network on the quality and reliability of the 4G network, where the optical core network is considered as the infrastructure of the 4G network, and any developments in the optical core network will enhance the capacities, and the availability of the 4G leased lines. The limitations in the leased lines of the 4G network can be listed as follows:

- The current structure of the optical system affects the quality and the availability of the 4G traffic in such a way that; There is no any types of correlation or association between the various domains of the optical topology, the connections between the working and protection routes between the different optical domains, the separate network managing systems (NMS's) for each vendor model, and the possible bandwidths in the various areas which are retained for its domain area only. With this model, it's challenging to implement the backup routes to the 4G services between the different areas domains in case of occurrence of any failure or crisis in one of these domains, also in the current long-distance OTN model which carry the traffic of 4G there is no any smart tools to put the performance and the latency of the backup routes in the considerations before choosing it for the restoration of the 4G traffic which may affects the customers of the 4G in a flapping way as a result of carrying the 4G traffic over backup links with high bit errors rate (BER).

- The high cost of the operations expenses (OPEX) in the OTN affects the 4G traffic in such a way that the operators tend to reduce the cost of building the 4G mobile network by reserving the leased lines with low capacities in the optical system to carry the mobile services of the customers between the different regions in the country, this will make congestions, low speed, and sometimes unavailability of the 4G network. The high cost of the operations expenses (OPEX) of the OTN as a result of there is no smart tools that can be used to optimize the operational costs in a robust optical network. The critical factors which affect the cost of the operations (OPEX) in the optical system are controlling the energy consumptions, reducing the time of the fault localization, monitoring the quality of the

transmission links in a periodic way and finally choosing the best routes for the creations of the new circuits.
- As a result of the 4G services is all IP traffic, the security of the client data of the 4G network is most vulnerable while it is traveling journey across the core optical network, where its security depends only on the standard encryption algorithms at the access layers. These signals travel through different layers in the optical core networks, and the operators of these networks rely mostly on the security algorithms of the client signals at the access layers only. The problem with the current security model of the 4G network is that the majority of the optical networks operators have no security algorithms at the physical layer of their optical system which is considered the weakest segment in all the communication network, and anyone can attack by splitting the optical signals and wiretapping it from any location of the network, as soon as the attacker reached to the optical signals of the OTN, he can retrieve the original data of the client signals of the 4G, the probability that the attacker can reach to original contents of the client data and break the encryption technique in the optical network is 100% success as a result of the standard encryption algorithms which are used in the OTN.

1.3 Objectives and Contributions of This book

The objectives and the outputs of the book are categorized into five items as following:

1. Distinguishing the main challenges of the 4G network and the optical core network.

Before proposing a new model of the mobile network, especially with the desire to implement the 5G technologies, the main challenges which affect the existing 4G mobile network should be studied. The study of the current difficulties in the cellular system is very critical to make any developments in the current communication model to be more productive.

2. Identifying the current model of the infrastructure of the 4G and the optical transport network.

The current model of the communication network has critical parts that affect the performance of the 4G network; one of these parts is the infrastructure of the 4G network, which is represented in the optical core network or the OTN. Defining the essential elements in the communication network is very important, where any developments in these few parts will affect in a significant way the performance of the mobile system and the availability of the smart applications inside the country.

3. Finding smart tools to reduce the cost of the operation of the Optical Core Network of the 4G.

The traditional way of performing the operational tasks in the core network is very costly and affects the investments in the mobile network. Proposing universal platform to control and perform the operational functions by using the advanced technologies of the artificial intelligence (AI), which can help in performing most of the routines tasks in the optical network smartly and automatically, will be considered as a revolution in the developments of the optical system and mobile services, This will reduce the cost of the operations by optimizing the used resources in the optical core network and will affect in a right way the performance of the 4G traffic.

4. **Proposing intelligent virtualization of the different optical network domains to enhance the resiliencies and the dynamic capacities for the 4G services.**

The structure of the current communication network makes the network domains, and the optical layers worked in isolated islands even though these domains are built in the same infrastructures of the optical core network or the region. Proposing a new model for slicing and virtualizing the different optical areas with work with each other's by managing and controlling all of them with one intelligent centralized controller, this will be done after rearranging the infrastructures of the optical network in clouding way which will help in the view of the complete picture of the several domains and will affect in a right way the availability and the resilience of the leased lines of the 4G network over the OTN.

5. **Proposing a new model for securing the 4G traffic and the services over the physical layer of the optical core network.**

Most of the security books were done on how to secure the 4G traffic in the access network only while few types of analysis were paid the attentions about the most penetrative part in the communication network which is the physical layer of the optical system, where the part of the network carries the traffic of the mobile network between the different regions inside the country. Proposing new cryptographic algorithms in the OTN will help to secure the traffic over all the parts of the communication network.

1.4 Book organization:

The book is planned as follows:
- Investigating the current challenges in the 4G network and the impact of the optical transport network. The discussion about the existing optical network structure and the weakest points in the construction of the system.
- Identifying the most penetrated parts in the communication network and reviewing the different studies about security algorithms in the optical system.
- Studying the difficulties of the traditional structure of the optical network and its impact on the developments and the stability of the mobile system, especially the resiliencies and the available capacities of the leased lines of the 4G network
- Studying the problems in the traditional ways of managing and controlling the operational tasks in the optical transport network which affect the cost of the leased lines of the 4G network
- Proposing new models for the security, the structure, and the control of the optical transport network by using the latest advanced technologies of the software-defined network and machine learning.

- Using SPSS and Neural Network Designer software to predict the parameters of the security model, the universal platform of the operation task, and the intelligent performance models from a dataset that is collected from a real long-distance optical network.
- Using practical experiments in the long-distance optical transport network to verify the effectiveness of the proposed models by slicing and virtualizing the different domains of the optical system and its impact on the resilience and the availability of the 4G traffic all over the country.

The book is organized into eight chapters, as follows:

- **Chapter 2** presents an overview of the advanced network technologies in our book, such as the software-defined network, machine learning, and the optical transport network.
- **Chapter 3** provides the literature review of the different challenges in the 4G network and the optical system and investigates the various studies to overcome these challenges.
- **Chapter 4** discusses the security of the client data signals over the optical transport network and provides the proposed security model in the OTN frames.
- **Chapter 5** discusses the tradition model of performing the operational tasks in the optical network and provides the proposed intelligent platform to execute these tasks in automatic methods.
- **Chapter 6** explains the challenges in the current structure of the long-distance optical network, illustrates the impacts of these challenges on the resilience and the availability of the 4G traffic, and introduces a proposed construction of the existing optical system by using the software-defined network and machine learning technologies.
- **Chapter 7** introduces the results and the analysis of the proposed models in chapters 4, 5, and 6.
- **Chapter 8** includes the conclusion and the future work of the book.

Chapter 2

The Advanced Technologies in the Optical Network

2.1 Introduction

This chapter reviews the advanced technologies in the optical network, which are promised to be used in the next generations of the communication markets by the telecom operators to transform their system to be smarter and more automated. In this chapter, we concentrate only on three advanced technologies in the optical network, which are used in our book, such as Optical Transport Network (OTN), Machine Learning technology (ML), and Software-Defined Network (SDN). The chapter is arranged as follows: part 2.2 provides an overview of the principles of optical transport technology, section 2.3 discusses the roles of the software-defined network in the optical system, and finally, section 2.4 provides the benefits of utilizing the machine learning technology in the optical system.

2.2 Optical Transport Network

The traditional communication networks were mainly based on the Synchronous Digital Hierarchy (SDH) as the standard of the international telecommunication union (ITU) G.707. The OTN is the latest transport technology for the Optical transmission Network, which is produced by the ITU G.709. The goal of the Optical Transport Hierarchy is to transfer a massive amount of data by using Optical Channel Data Units (ODUs) over the optical network. The Optical Transport Hierarchy (OTH) Architecture formed for the Optical Transport Network (OTN) and its structures represent two interface types: Inter-domain interface and Intra-domain interface. The OTN Inter-domain interfaces are defined with 3R processing at every edge of the interface, and this would be the interface among various Operators. It could also be considered as the interface among multiple Vendors within the same Operator, as shown in figure 2.1. As shown in figure 2.2, G.709 defines the several layers of the OTN hierarchy. The OTH provides more than three multiplex panels k, including k=1, 2, and 3, and represents the similar Optical Data Units (ODUk). Every ODUk has an overhead range plus a payload range. The Optical Payload Units (OPUk) encapsulates the Customer signal such as the SDH data and performs any required justification, and It is similar to the Path layer in the SDH, that is mapped at the beginning stations, de-mapped at the sink stations, and does not changing by the network. The ODUk plays related purposes as the Line Overhead in the SDH. The Optical Transport Units (OTUk) includes the Forward Error Correction (FEC) and performs associated objects as the Section Overhead in SDH. Following the FEC is determined, the signal is then sent to Serializer - Deserializer to be turned to the optical domain. The data rates were designed so that it could transfer different signals types such as SDH signal efficiently. The bit rates of the OTH are shown in the following tables 2.1, 2.2, 2.3, and 2.4.

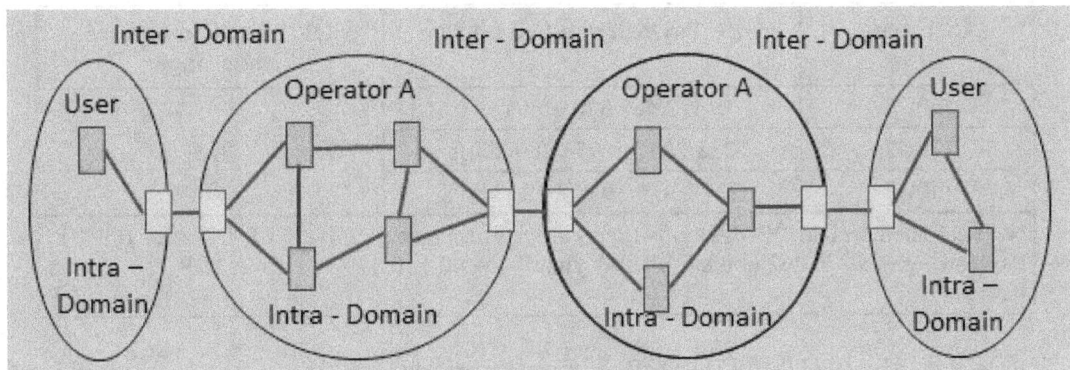

Fig. 2.1. The OTN interfaces within the Operator or a Vendor domain

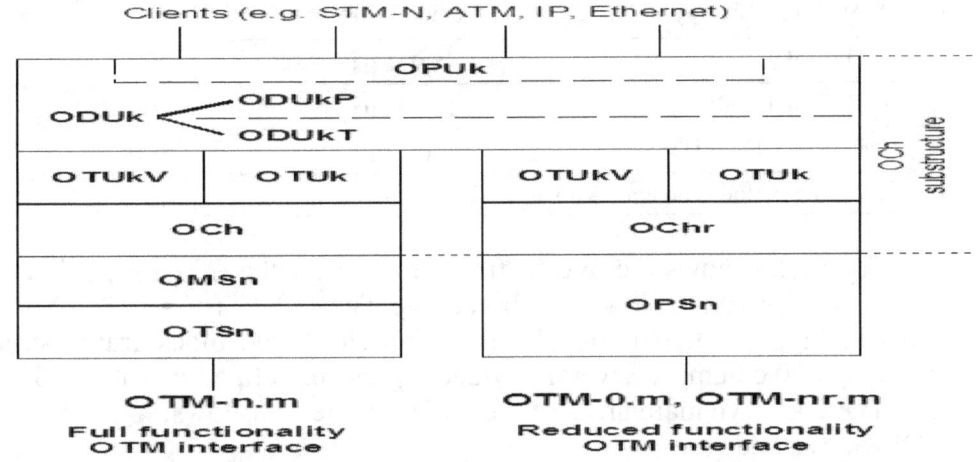

Fig. 2.2. The mapping of the client data to be used in the OTN

Table 2.1: The OTU types and capacities

OTU type	OTU nominal bit rate	OTU bit rate tolerance
OTU1	255/238 * 2 488 320 kbit/s	± 20 ppm
OTU2	255/237 * 9 953 280 kbit/s	
OTU3	255/236 * 39 813 120 kbit/s	
NOTE - The nominal OTUk rates are approximately: 2 666 057.143 kbit/s (OTU1), 10 709 225.316 kbit/s (OTU2) and 43 018 413.559 kbit/s (OTU3).		

Table 2.2: The ODU types and capacities

ODU type	ODU nominal bit rate	ODU bit rate tolerance
ODU1	239/238 * 2 488 320 kbit/s	± 20 ppm
ODU2	239/237 * 9 953 280 kbit/s	
ODU3	239/236 * 39 813 120 kbit/s	
NOTE - The nominal ODUk rates are approximately: 2 498 775.126 kbit/s (ODU1), 10 037 273.924 kbit/s (ODU2) and 40 319 218.983 kbit/s (ODU3).		

Table 2.3: The OPU types and capacities

OPU type	OPU Payload nominal bit rate	OPU Payload bit rate tolerance
OPU1	2 488 320 kbit/s	± 20 ppm
OPU2	238/237 * 9 953 280 kbit/s	
OPU3	238/236 * 39 813 120 kbit/s	
NOTE - The nominal OPUk Payload rates are approximately: 2 488 320.000 kbit/s (OPU1 Payload), 9 995 276.962 kbit/s (OPU2 Payload) and 40 150 519.322 kbit/s (OPU3 Payload).		

Table 2.4. The OTUk/ODUk/OPUk frame periods

OTU/ODU/OPU type	Period (note)
OTU1/ODU1/OPU1	48.971 µs
OTU2/ODU2/OPU2	12.191 µs
OTU3/ODU3/OPU3	3.035 µs
NOTE - The period is an approximated value, rounded to 3 digits.	

Figure 2.3 shows the overall frame format for the OTUk signal. The various fields are explained in the following subsections. The OPUk (k = 1, 2, 3) frame structure is shown in Figure 2.3. It is organized in an octet-based block frame structure with four rows and 3810 columns. Several Payload Types are defined in Table 2.5.

There are two main areas of the OPUk frame as follows:

OPUk overhead area

OPUk payload area

Mapping SONET/SDH into OPUk both synchronously or asynchronously is the common mapping technique. Synchronous mapping is a subset of asynchronous mapping, as shown in figure 2.4, and 2.5 the ODUk frame structure. It is arranged in an octet-based block frame construction by four rows plus 3824 columns. There are three main parts of the OTH frames are included in the OTUk, ODUk, and OPUk. Where columns 1 to 14 of rows 2-4 is assigned to ODUk overhead, columns 1 to 14 of row 1 is reserved for frame alignment in the OTUk overhead, and columns 15 to 3824 of the ODUk frames are assigned to OPUk part. The information on the overhead in the ODUk is added to the data payload to produce an ODUk. It covers data concerning maintenance and operational duties to maintain optical channels.

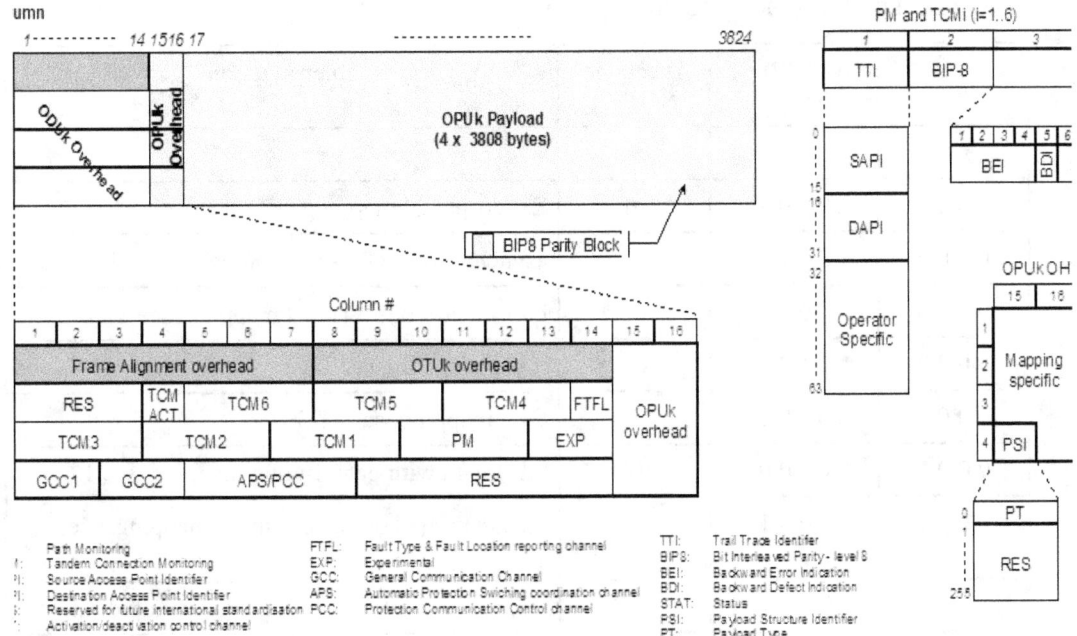

Fig. 2.3. The OTN Frame Format

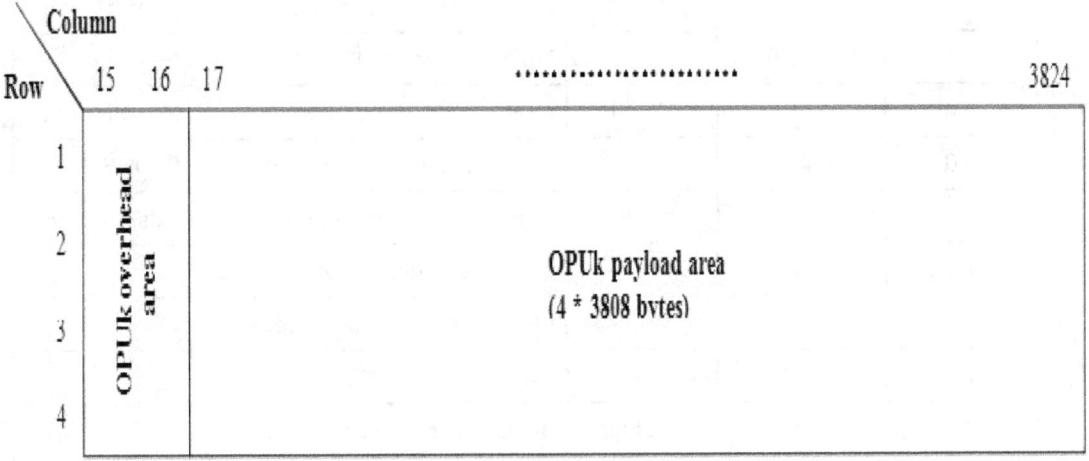

Fig. 2.4 The OPUk frame structure

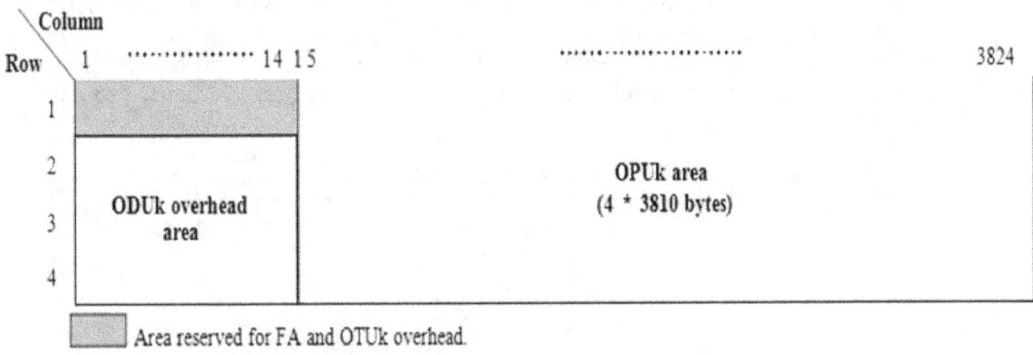

Fig. 2.5. The ODUk frame structure

Table 2.5: The Payload type code points

MSB 1 2 3 4	LSB 5 6 7 8	Hex code (NOTE 1)	Interpretation
0 0 0 0	0 0 0 1	01	Experimental mapping (NOTE 3)
0 0 0 0	0 0 1 0	02	asynchronous STM-N mapping, see 17.1
0 0 0 0	0 0 1 1	03	bit synchronous STM-N mapping, see 17.1
0 0 0 0	0 1 0 0	04	ATM mapping, see 17.2
0 0 0 0	0 1 0 1	05	GFP mapping, see 17.3
0 0 0 1	0 0 0 0	10	bit stream with octet timing mapping, see 17.5.1
0 0 0 1	0 0 0 1	11	bit stream without octet timing mapping, see 17.5.2
		12 - 54	reserved for future international standardisation
0 1 0 1	0 1 0 1	55	Not available (NOTE 2)
		56 - 65	reserved for future international standardisation
0 1 1 0	0 1 1 0	66	Not available (NOTE 2)
		67 - 7F	reserved for future international standardisation
1 0 0 0	x x x x	80 - 8F	reserved codes for proprietary use (NOTE 4)
		90 - FC	reserved for future international standardisation
1 1 1 1	1 1 0 1	FD	NULL test signal mapping, see 17.4.1
1 1 1 1	1 1 1 0	FE	PRBS test signal mapping, see 17.4.2
1 1 1 1	1 1 1 1	FF	Not available (NOTE 2)

NOTE 1 - There are 228 spare codes left for future international standardisation.

NOTE 2 - These values are excluded from the set of available code points. These bit patterns are present in ODUk maintenance signals.

NOTE 3 - Value "01" is only to be used in cases where a mapping code is not defined in the above table. By using this code, the development (experimental) activities do not impact the OTN network. There is no forward compatibility if a specific payload type is assigned later. If a new code is assigned, equipment using this code shall be reconfigured to use the new code.

NOTE 4 - These 16 code values will not be subject to standardization.

The ODUk overheads consist of parts assigned to the end-to-end ODUk path and six levels of tandem connection monitoring. The ODUk path overhead is eliminated wherever the ODUk is assembled and disassembled. The tandem connection (TC) overhead is joined and ended at the source and sink of the analogous tandem connections, sequentially. The ODUk overhead position is shown in Figure 2.6. In the OPUk Columns, 15 to 16 are assigned to OPUk overhead range, where Columns 17 to 3824 are attached to the OPUk payload range. The OPUk overhead data is combined with the OPUk data payload to produce an OPUk. It carries data to assist the adjustment of client signals. The eliminations of The OPUk overhead are done where the OPUk is assembled and disassembled. Figure 2.7 shows the overhead bytes of the OPUk.

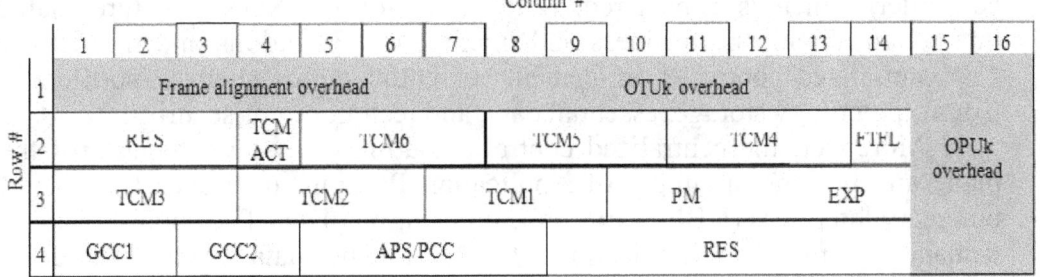

Fig.2.6. The ODUk overhead position in the frame structure

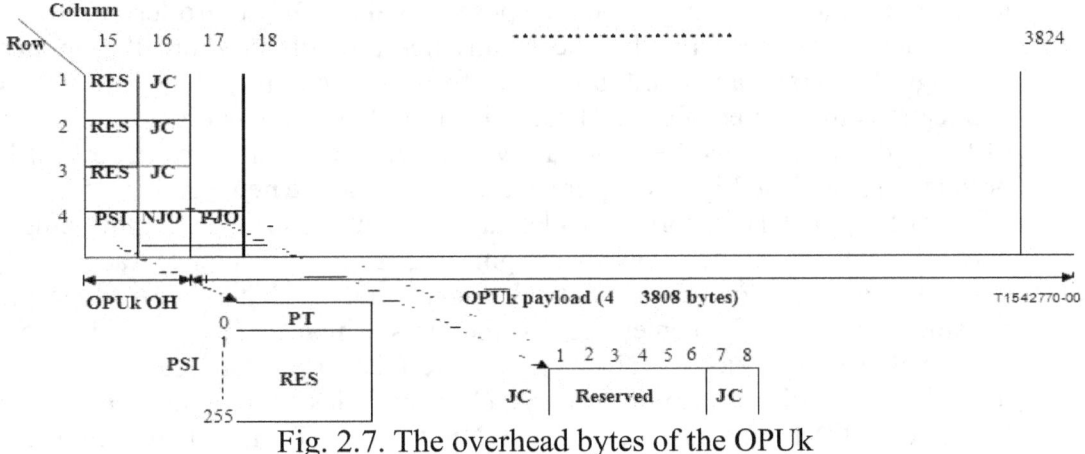

Fig. 2.7. The overhead bytes of the OPUk

2.3 Software-Defined Network Fundamentals

The software-defined network (SDN) is a promising technology concerning network planning, structure, and control. The fundamental approach is to separate the decoupling of the control plane from the data plane. The SDN is developing the perspective of today's networking society and is pushing disruptive creativity in every networking area such as data center, packet switching, enterprise network, wireless access network, clouds, and finally, optical transport networks. Commonly, SDN is formed of three principal planes with unique functionalities and interfaces, as shown in figure 2.8 [4].

The Data Plane is the base of the SDN structure; it is accountable for controlling packets in the data route based on rules derived from the SDN Control Plane. The data plane is formed through the processing in the network elements (NE's) such as router and optical transmission equipment with the data traffic in the network.

While in a traditional network, the control and the data planes are performed in the firmware of Network Elements, while in the SDN, the control duties are separated from the NE's, which does not make any complicated distributed algorithms.

The data plane transmits slips and adjusts packets depending on orders from the control plane through the south boundary Interface (SBI). The Control Plane has performed as a standalone software model and separated from the firmware of NE's. The standalone software model enables the availability of the network resources to be programmable access, delivering network rules, and performs the controlling duties of the network management. The SDN control plane is designed by using a centralized controller, which is considered as the brain of the SDN structure that manages the interaction among the business and intelligent applications and the network Elements. The centralized controller implements essential duties such as notifications and NE's Control, topology storage, essential data, and techniques of security.

Moreover, the centralized controller outlines low-level aspects of the forwarding plane and presents a supply of Application Program interfaces (APIs) with the North boundary interfaces (NBI - API) to the application plane. The control plane transposes the demands from SDN applications falling to the data plane by administering the configuration of forwarding components utilizing the south boundary interfaces. The application Plane determines the network control parameters; the applications get a clear network illustration from the SDN controller, and based on this; execute the control-logic to create conclusions that will be transposed via the SDN controller into rules to program the resources of the network. The creativities are offered from the application plane, although the commands are produced from the control plane, the broad range of applications in different fields. The application plane depicts one of the first appearances of the SDN as it allows the opportunity to reproduce regulations in several languages to produce industry intelligence, optimization, and produce new services [5].

The Optical Transport networks have characteristics usually not being in computer networks wherever the SDN paradigm began, like resilience, advanced structures, heterogeneous topologies, and optical domain failure that require to be practiced into consideration while implementing the thoughts which were proposed by SDN. The T-SDN includes expansions to abstractions, interfaces, rules, and control plane components to collaborate with transport network characteristics and to master the constraints of OpenFlow (ONF) in this area. The ONF Optical Transport Working Group (ONF-OTWG) redefined as the Open Transport Working Group offered whatever they called OpenFlow-enabled Transport SDN structure, which is mainly based on OpenFlow [7].

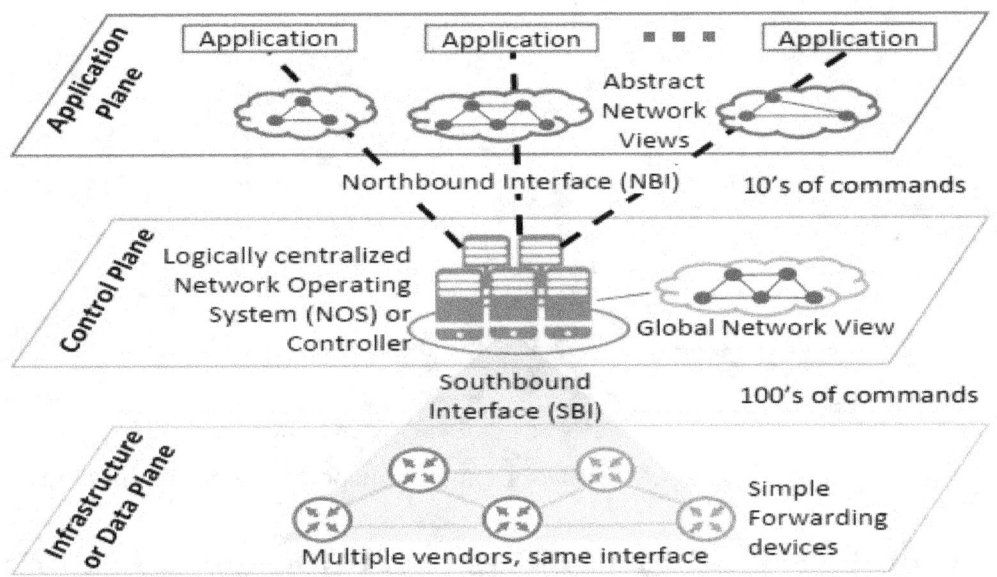

Fig. 2.8. The SDN Model Architecture

In 2013 the ONF announced the OpenFlow Switch characteristics (v1.4.0) that included for the first time for the assistance to the optical interfaces. Although, the industry is yet in development in ONF-OTWG to determine reliable and standard NBI and SBI designation for SDN/OpenFlow-based T-SDN, including expansions for supervision and control of the OTN, and wireless transportation. This effort concentrates on the optical transport network, rather than in the wireless transportation, which is a field of a large business with the appearance of 5G mobile networks, the Internet of things and mobile-cloud period. To cope with such complicated optical network structures which rely on a vendor based Network Management System (NMS) , furthermore makes light path's computation, optical resource allocation, optical performance monitoring, and NE's configurations, as shown in Figure 2.9 the existing optical networks perform the GMPLS protocol series as the spread control plane for dynamic route setup [7, 8].

The NMS mutually with GMPLS provides a "great switch" concept that covers the topology and the optical complexity to the Operation Support System (OSS) and applications. The optical network devices providers have expanded their solutions' competing advantages via offering exclusive technologies with unique features and developing their NMS's. This action directed to different data planes, with interoperability concerns with different vendors' equipment. As a result, the transportation network of operators is formed through administratively private vendor islands, each managed by a centralized NMS. This heterogeneous situation depicts a significant difficulty in determining common concepts for T-SDN and finding complete visibility and authority across the multi-vendor, multi-layer, and multi-domain.

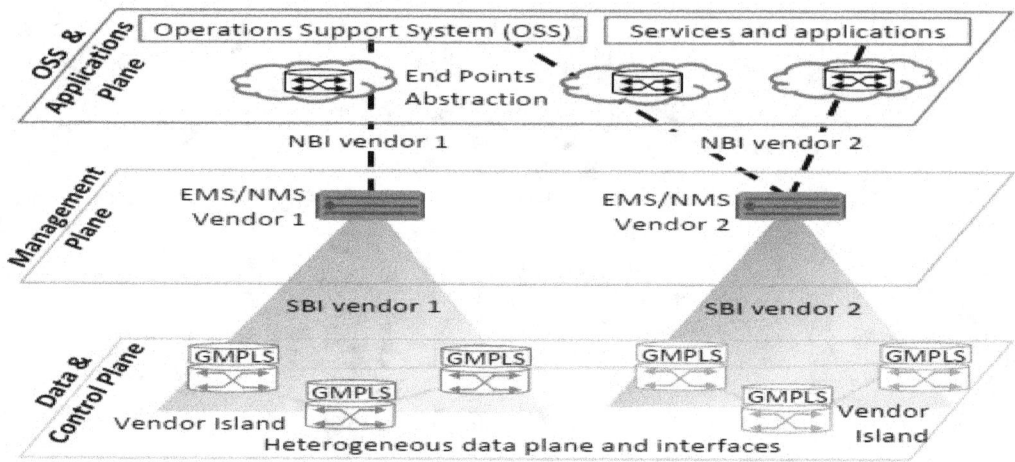

Fig. 2.9. The SDN Legacy transport network architecture.

2.4 The Machine Learning Technology in the Optical Network

Machine Learning (ML) is an outstanding precise system that permits computer methods to answer the difficulties by imitates the complicated biological rules such as learning, rationalizing, plus self-correction. The ML algorithms are classified into three principal categories, which are unsupervised learning, supervised learning, and reinforcement learning, as represented in figure 2.10. The algorithms of the ML have been actively implemented in a wide variety of challenges.

Supervised learning is utilized in a variety of applications, such as spam discovery, speech identification, and object verification. The purpose is to predict the output variable values after defining the value of the input variables. The output variable can be discrete or continuous.

A training data set includes N examples of the input variables and the analogous output values. Various learning techniques create a function that permits predicting the value of the output variables in agreement to a distinct value of the inputs. Supervised learning can be split down into two principal categories explained as parametric forms, wherever the number of parameters to utilize in the model is fixed, and nonparametric forms, where their number is reliant on the training set.

In the parametric models, the output function is a mixture of a determined quantity of parametric base parties. These models utilize training data to evaluate a proposed set of parameters. Following the learning step, the training data can be dismissed since the forecast in agreement to further inputs is calculated utilizing just the learned parameters. Linear representations for regression and organization, which consist of a linear mixture of fixed nonlinear functions.

In the Nonparametric models, the character of parameters relies on the training set by retaining a subset of the training data and utilizes them during prediction. The most utilized methods in this model are the k-nearest neighbor models and the support vector machines (SVM), where both methods can be employed for regression and analysis obstacles. In the situation of k-nearest neighbor systems, aggregate training data examples are saved. While prediction, the k-nearest cases to the new input value are recovered. For an analysis obstacle, a deciding mechanism is used for regression difficulties.

In the SVM, base functions are focused on training examples; the training method chooses a subset of the base functions number of selected basis functions, and the number of training examples that have to be saved, is typically extremely smaller than the cardinality of the training data set. The SVM builds a linear choice boundary with the most significant potential distance from the training examples. Only the nearest points to the separators and the support vectors are stored [9].

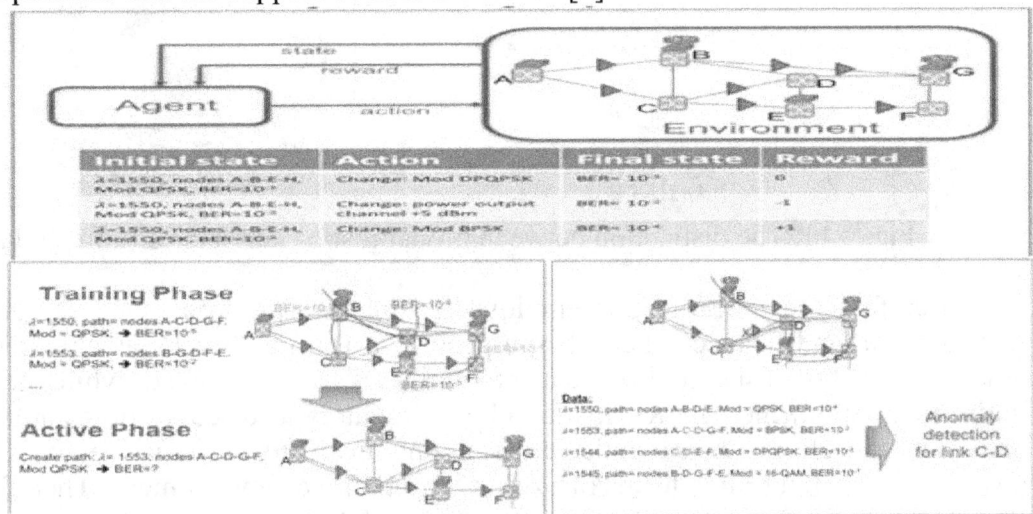

Fig. 2.10. The Categories of the ML algorithms

In unsupervised learning, the training dataset consists just of a collection of input vectors. While unsupervised learning can direct various tasks, clustering analysis is the usual common. Clustering is the method of grouping data so that the intra-cluster correlation is high, whenever the inter-cluster relationship is low. The association is typically represented as a distance function, which relies on the kind of data. There is a variation of clustering methods such as Gaussian mixture and k-means models. k-means is possibly the usual popular clustering algorithm [10]. It is considered as an iterative algorithm originating with a primary partition of the data within k clusters. Then the middle of every cluster is calculated, and data points are attached to the cluster with the nearest center. The method of center calculation and data designation is returned until the classification does not change or a predefined highest number of repetitions is passed. Performing such an algorithm may stop at a local best partition.

Furthermore, k-means is adequately acknowledged to be sensible to outliers. It is worth remarking that there are techniques to calculate k automatically [11]. While k-means specify every point uniquely to an individual cluster, probabilistic strategies provide a flexible appointment and give a degree of the uncertainty correlated with the designation. Figure 2.11 displays the distinction between probabilistic Gaussian Mixture and k-means models. A probabilistic Gaussian Mixture Model is a linear superposition of Gaussian distributions, and the common generally accepted probabilistic methods to clustering. The parameters of the technique are the mixing coefficient of the mean, each Gaussian component, and the covariance of each Gaussian distribution.

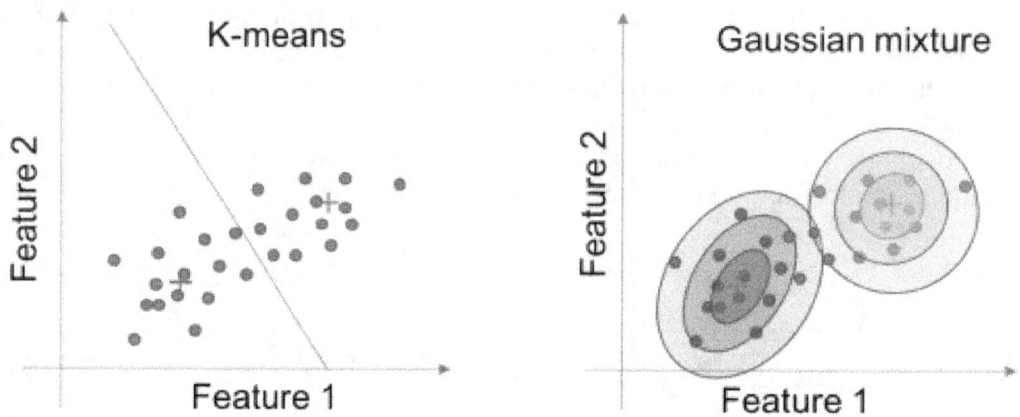

Fig. 2.11. The distinction between probabilistic Gaussian Mixture and k-means

Reinforcement Learning is employed to manage purposes such as robotics, inventory management, and finance, where the purpose is to learn a strategy as an example of mapping among the status of the environment into activities to be achieved, while straight associating with the environment. The RL model allows operators to discover by examining the possible activities and refining their practice utilizing only evaluative criteria feedback, related to as the award. The operator aims to maximize its long-term accomplishment. Therefore, the operator does not only bring into account the instant reward, but it estimates the outcomes of its actions in the future. Postponed reward and trial-and-error establish the two most notable features of RL. Reinforcement Learning is usually done in the circumstances of Markov decision processes (MDP). Many of the applications can be made by ML in the different fields, as shown in figure 13, and there are several use cases that can help in the employment of the Machine Learning and the data analytics methods in the physical and network layers of the optical transport network use cases. In the Physical layer area, many challenges require to be discussed, typically to estimate the performance of the transportation system and to investigate if any degradation impacts surviving light paths. The idea of Quality of Transmission usually refers to several physical layer parameters, such as OSNR, Bit Error Rate (BER), Q-factor, which influence on the availability at the receiver of the optical signal.

Such parameters provide a quantitative action to check if a decided level of quality of transmission would be confirmed and are influenced by many tunable configuration parameters, such as modulation form, coding rate, baud rate, and physical path in the network. Consequently, optimizing this opportunity is not little, and usually, this wide variation of probable parameters challenges the experience of a method engineer to direct all the probable mixtures of the light path deployment. Another use case of ML that can be implemented in the optical networks is light path provisioning to be more dynamic. Unfortunately, dynamic structure and tear-down of light paths across various wavelengths drive the operators to reconfigure network elements on the fly to keep the stability of the physical layer.

In reply f this development, the Erbium-Doped Fiber Amplifier (EDFA) flag from wavelength-dependent power journeys. Precisely, while a new light path is placed or during a current light path is cut down, the variance of signal power levels between various channels depends on the particular wavelength being added-dropped into or from the system.

Therefore, automated administration of amplification of the signal power levels is wanted, particularly in the situation of a cascade of many EDFAs is crossed to avoid that extreme amplification power inconsistency. In the network layer, there are several use cases for Machine Learning in the optical network that can be implemented, such as the traffic forecast in the time domain, which enables operators to plan and manage their networks efficiently. In the planning phase, the traffic forecast permits to overcome the extra provisioning as much as possible. Throughout network administration, resource utilization can be optimized by implementing traffic handling based on ultimately rerouting the current traffic, the real-time data, and the reserving of the required resources in the future for the incoming traffic demands. Another use case in the network layer is the Provisioning of the installed light paths or repair of the current services in any failure in the network, which demands complicated and quick determinations that rely on many fast-evolving data. Figure 3.12 shows the different applications of the AI in the optical network.

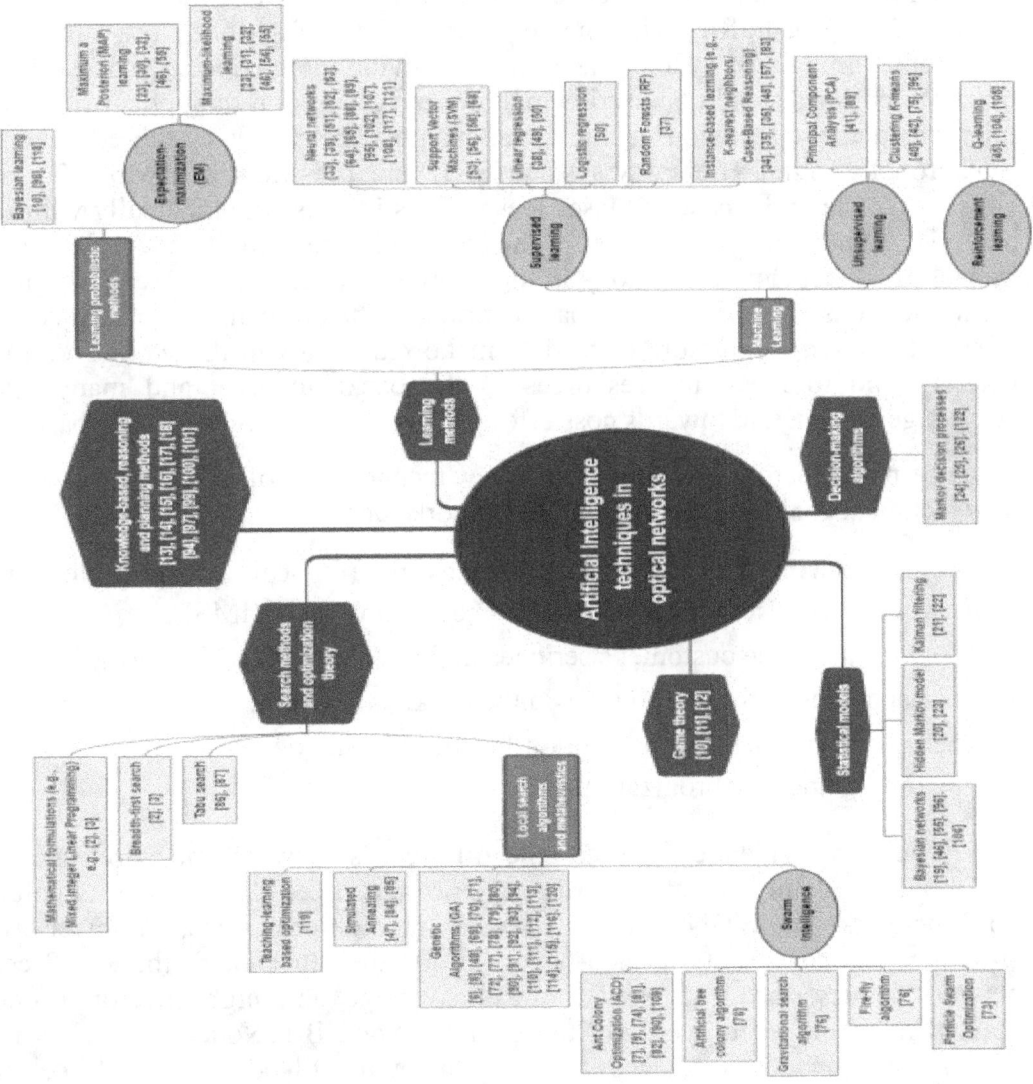

Fig. 2.12. The different applications of the AI in the optical network

Chapter 3

Review of the Major Challenges in the backhaul of the 4G Network

3.1. Introduction

The connectivity of the backhaul in the mobile network depends mostly on three types of transmission media; copper, microwave, and optical fiber. The copper lines provide connections with speeds of 2.048 Mb/s (E1) or 34Mb/s (E3). The available leased lines of copper media dominate the backhaul solutions of 4G Networks, as it provides proper support for voice traffic, with the required quality of service, low latency, low delay variations, timing, and synchronization. To provide more capacities and different speeds other leased lines should be used, which depend mainly on the optical fiber cables or the microwave as transmission media, as an example the speeds of these leased lines in the Synchronous Digital Hierarchy system (SDH) are 155 Mb/s, 622 Mb/s, 2.5 Gb/s and 10 Gb/s, at the same the cost of theses leased lines increases dramatically with its capacity, and this is not an efficient choice for the backhaul in the mobile network. Due to the vital impact of backhauling technology in the mobile network, some researchers have recently mentioned it as the "Telecom Global Warming." The problem of the high cost of the least lines caused some operators to tend to make reductions in the workforce in the access network and to move the resources to the backhaul area, and many operators are encouraged to migrate towards cost-effective backhaul solutions packet-based [12].

It can be noted that the migration to the new technologies of 4G (LTE) network raises new technical challenges in the mobile network such as:

- Integration with other heterogeneous networks (Optical Transport Networks, IP core network, Wi-Fi Network, and IP multimedia subsystem (IMS).
- The security of the customer's services in the Optical Core Networks.
- The capacities of the leased lines in the backhaul section.
- The cost of the energy consumption for the base station and the optical network.
- The timing and synchronization of IP traffic.

The first challenge discusses the differences between the wireless data network (WDN) and the 3rd generation partnership project of cellular long-term evolution (3GPP LTE) network. The WDN is characterized by its high peak rates, but lower efficiency for small packets, and by its limited coverage. On the other hand, the 3GPP cellular LTE network is described by its widespread coverage and high spectral efficiency. The placement of the radio access in the evolved Node B (eNodeB) elements makes them vulnerable to unauthorized access because the evolved Node is located in an unobserved place. Further, internetworking with heterogeneous optical transport networks shows up the vulnerability of these networks to the direct external threats and carries a gap trap for LTE security [13].

The first challenge discusses the differences between the wireless data network (WDN) and the 3rd generation partnership project of cellular long-term evolution (3GPP LTE) network. The WDN is characterized by its high peak rates, but lower efficiency for small packets, and by its limited coverage. On the other hand, the 3GPP cellular LTE network is described by its widespread coverage and high spectral efficiency. The placement of the radio access in the evolved Node B (eNodeB) elements makes them vulnerable to unauthorized access because the evolved Node is located in an unobserved place. Further, internetworking with heterogeneous optical transport networks shows up the vulnerability of these networks to the direct external threats and carries a gap trap for LTE security [13].

Due to the contradiction between tremendous expansion of mobile data traffic and the restricted wireless spectral resources at traditional RF bands for both cellular and Wi-Fi networks, more aggressive spectral reuse by reducing the cell size, new spectral consideration at higher RF bands, and collaborative multipoint operation among the remote radio heads (RRHs) are the three main trends for high-capacity and high capacity wireless access networks, and these techniques improve the total system capacity. To support small cells and next-generation Wi-Fi LAN networks, the optical fibers rather than copper cables are considered as ideal backhaul and front-haul media to provide enormous bandwidth and future-capacity upgrade. Therefore, integrated optical and wireless access technologies for the next-generation Wi-Fi and 5G wireless communications co-design of the optical to electrical and air interface become important topics shortly. Many factors affect the technologies for next-generation Wi-Fi, 5G small-cell systems, and WDM-PON will be discussed in this survey.

The energy consumption of the LTE-based radio access network is affected by the architecture of the network between the remote radio heads (RRHs) and base stations (BSs). There are two essentials different in the implementation model of the LTE network; a decentralized model where base stations (BSs) act all required tasks of radio access network (RAN), and a centralized model where the majority of the required tasks of radio access network (RAN), is performed by central baseband units (BBUs). In the decentralized model, the base stations (BSs) are liked via backhaul (BH) to the core network as well as to each other, in the centralized model; the remote radio heads (RRHs) only transform the analog signal to the digital domain and forward it to centralized central baseband units (BBU). The synchronization is the essential request in many packet-based networks and real-time application domains, such as the internet of things (IOT), automated medical systems, and smart solar applications. In this paper, a study was done for the synchronization standards of packet-switched networks. Further, this survey reviews the recently proposed techniques to improve synchronization accuracy and overviews of the major applications needing synchronization and their requirements.

3.2 The Challenges of 4G Network

We could summarize the significant challenges of LTE networks as follows:
- Integration of LTE Network with other IP core networks
- Security in backhaul and optical transport network of the LTE Network
- Available Capacities in the backhaul section of LTE Network
- Energy consumptions in LTE 4G Network and the backbone optical network
- Timing and synchronization of LTE Network

In the following subsection, the challenges are explained in detail.

3.2.1. Integration of the LTE Network with other IP Networks:

Due to the high speed of data rates of 4G (LTE), the customers of mobile services are exceeded 1.3 billion by the end of 2018 [4]. The existing design of the LTE network and its relations with the other networks (Wi-Fi & IMS & IP-Core Network and OTN) make the cost per bit of 4G very high, which affects the performance of the network.

3.2.1.1. Integration with IP Core Networks:

The implementation of the mobile network contains two layers as shown in figure 3.1 : the physical layer (L2) which provides the required connectivity and transport functionality to the logical layers from and the access network, and consists mainly of the evolved Node B (eNodeB) that provides the radio access to the user equipment (UE), and logical layer (L3) consists of many network elements, These network elements perform the mobility and transport data from mobile devices across the mobile network. The backhaul contains all of the needed aggregation switches to forward the traffic from the access network to the core network, which perform the mobility and billing tasks such as: Mobility Management Entity (MME), Serving Gateway (SGW), Policy and Charging Rules Function (PCRF), and Home Subscriber Server (HSS) [14]. The integration in the LTE networks is performed through different methods depending on whether the new target eNodeB is under different Tracking Area ID (TAI) and associated with the same Mobility Management Entity (MME) or the different Tracking Area ID (TAI) belongs to other MME. In the first scenario, the mobility is done by using the protocol S1-MME interface, which defined between the evolved Node B (eNodeB) and the Mobility Management Entity (MME). The mobility in the 2'nd scenario is done by using the X2 interface protocol, which is defined between evolved Node Bs (eNodeBs). Figure 3.2 shows the control functionality for managing the handover operation according to the S1-MME interface between logical elements (e.g., eNodeB, MME, and SGW) when the Mobility Management Entity change is required [15].

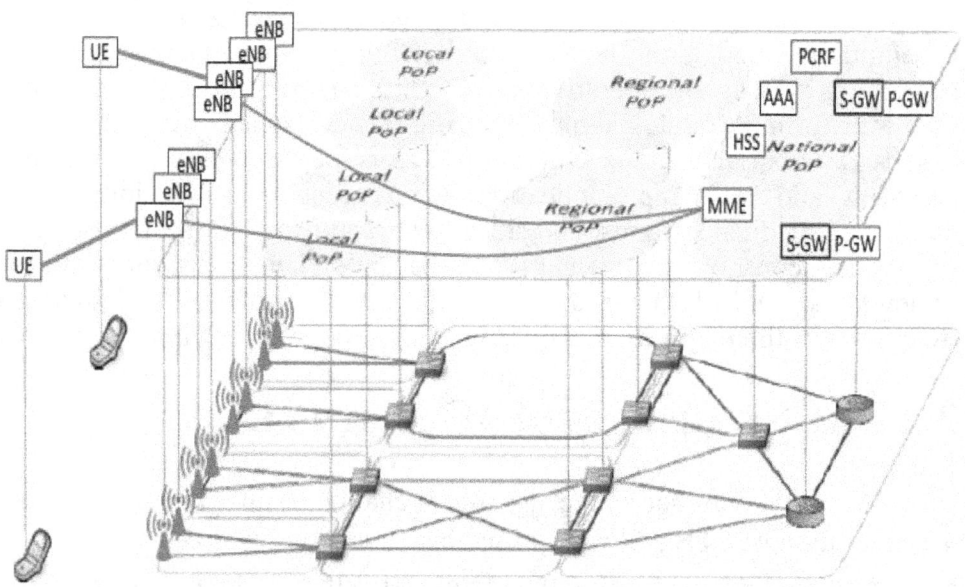

Fig. 3.1. The physical and logical layers in the mobile network

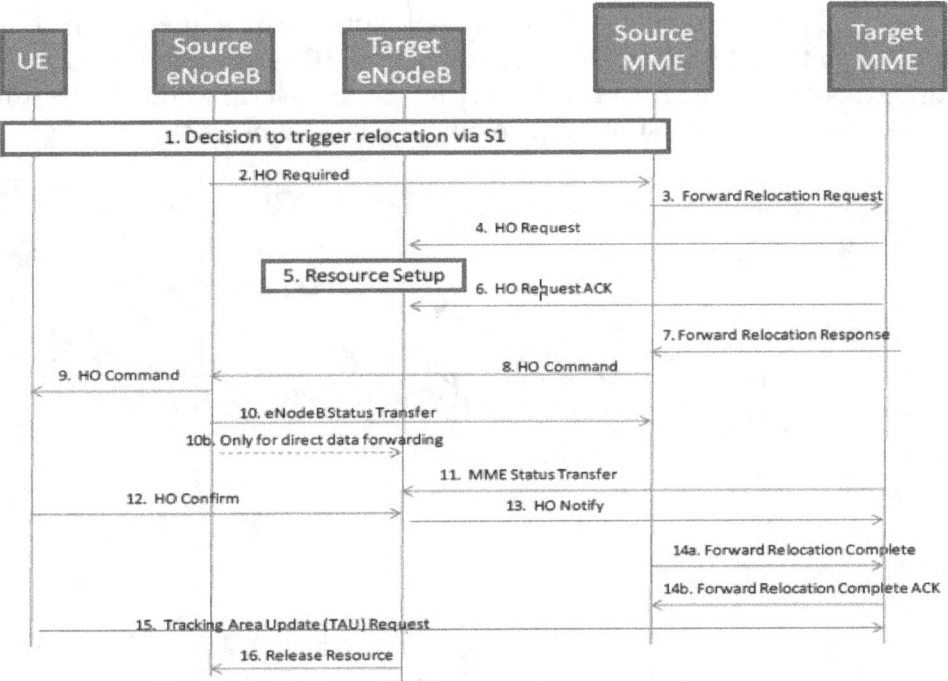

Fig. 3.2. The Logical Elements and Control Process

From the mobility point of view, the critical challenge in the integration operation of the IP protocol is the IP address, which identifies the node and fixes its location to a specific IP subnet. The solution in the mobile network consists of using tunneling for user equipment (UE). General packet Radio Service (GPRS) Tunneling Protocol (GTP) tunnels and assigns uniquely IP addresses that are established between evolved Node B (eNodeB) and service Gateways (SGW), and identify traffic flows that receive a standard quality of service treatment between a user equipment (UE) and packet data network gateway (PDN-GW). The traffic flow template (TFT) is used for mapping traffic to an evolved packet system (EPS) Bearer. The General packet Tunneling Protocol (GTP) tunnels the endpoint identifiers, which are included in each user data packet and separates the end-users identifiers (TEID) and its bearers of a particular user as shown in figure 3.3. While the user equipment (UE) proceed to a new evolved node B (eNodeB), the General packet Tunneling Protocol (GTP) has to be recreated between the new evolved node B (eNodeB) and the Serving Gateway (SGW), at the same time the inner data flow keeps the original IP address. The S1 interface initiates and manages the handover process, as shown in figure 3.4. The Mobility Management Entity (MME) is aware of the mobility process and communicates with the Serving Gateway (SGW) to recreate the General packet Tunneling Protocol GTP tunnel between the new eNodeB and the SGW [15].

Many studies were done on the integration of the LTE network with the different IP Networks, and most of them proposed a flatter architecture of the LTE Network together with applying the software-defined network (SDN) by integrating MME with controller functionality. The control plane of the mobile network is simplified by merging the LTE network elements such as MME, SGW into a centralized network component.

This new centralized network element that includes the SDN controller will be virtualized, and multiple instances can be installed in the data centers to manage the simplified data plane. As a result, a new model in the LTE networks will be raised, where the transportation of any mobile data is simplified and depends mostly on the SDN switches.

The proposed model of SDN based flatter architecture with LTE network elements will be suitable for cloud radio access network (CRAN), and this model will improve the throughput of the LTE Network by reducing the overhead (GTP-U) from the user plane between eNodeB and SGW and voiding fragmentation as shown n figure 3.5 [16].

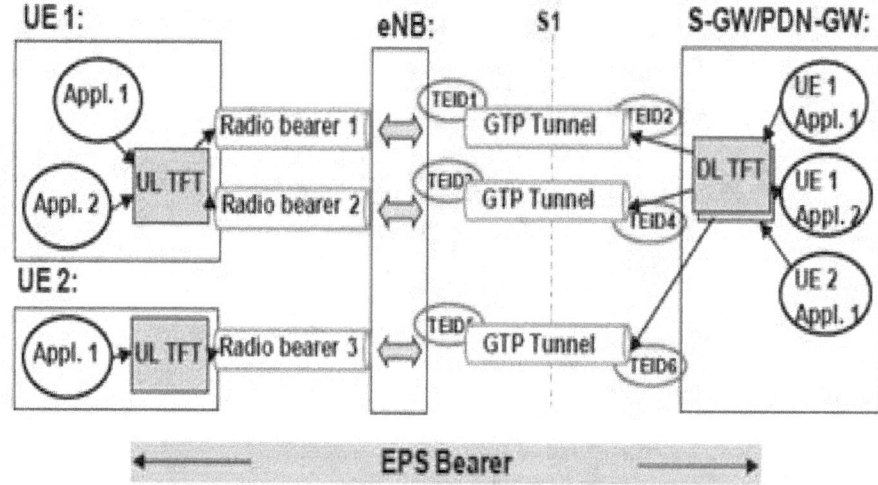

Fig. 3.3. The End-Users Barriers and Mapping to GTP

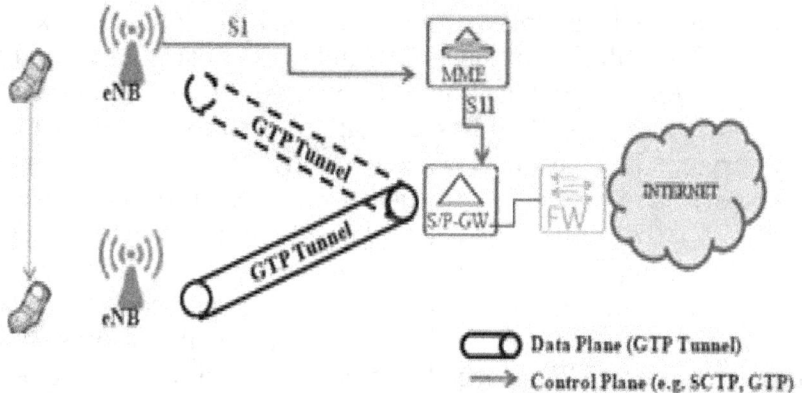

Fig. 3.4. The Handover Process from MME through S1

3.1.2. Integration with Wireless LAN

In the traditional approaches, the integration with Wi-Fi makes the operators offload the traffic from Wi-Fi while they are providing complete control and visibility of the user situation, as shown in figure 3.6.

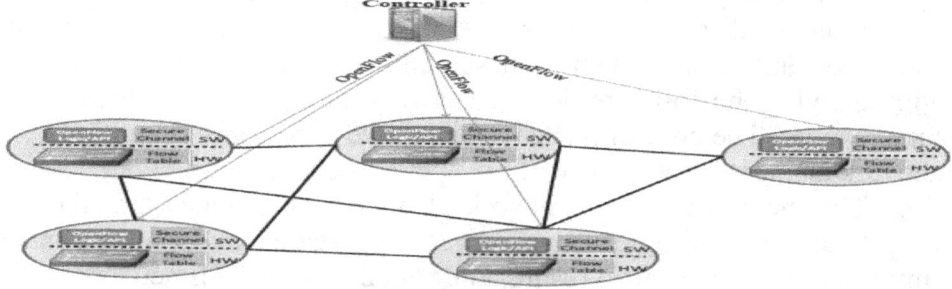

Fig. 3.5. The SDN Based Network

The Mobile network and the Wi-Fi integration models can be sorted into two different techniques [15]:
- Managed Offload: Integration with core network authentication and policy management systems
- Integrated Offload: Integration of Wi-Fi data traffic into the core network for mobility and offload the traffic through the core or the access network.

The 3GPP has standardized three main models [17]:
- The local IP access (LIPA) technique transfers data in a local network.
- The selected IP traffic offload (SIPTO) technique offloads the traffic to another network to reduce the load on the core network.
- IP flow mobility (IFOM) offers dynamic offloading between LTE and WLAN to balance the load of Radio Access Networks (RANs).

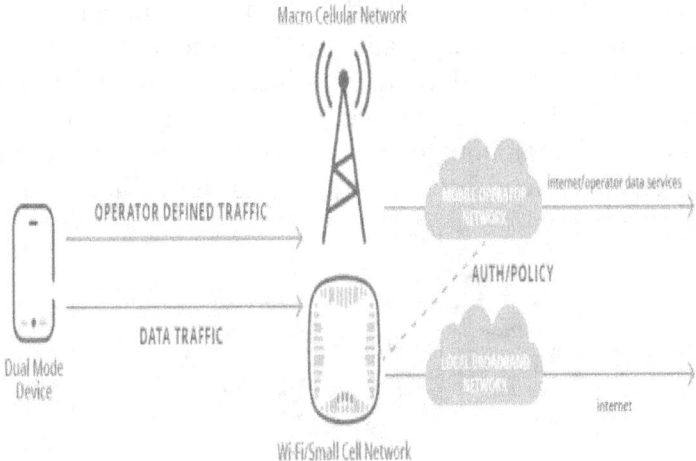

Fig. 3.6. The Architecture of Wi-Fi with LTE Network

The mobile network (MN) transmits data via LTE and WLAN links jointly. The home agent (HA) identifies a home of address (HoA) to the LTE link and another care-of address (CoA) to the WLAN link. The binding update (BU) of the MIPv6 protocol is added with the mobility option (CoA and BID) to support the offloading of multiple IP flows. The binding identification (BID) is the unique identification of an IP flow, while the care-of address (CoA) identifies Radio Access Networks (RANs). The home agent (HA) manages the IP flows between the mobile network (MN) and the corresponding node (CN) according to parameters such as flow identifier (FID), traffic selector, and FID priority [18].

In Wi-Fi/LTE heterogeneous networks, current algorithms for integration allow users access to one or another or concurrent use of both radio interfaces, but for different applications. For example, exploring the web site on the internet, the user may be switched from LTE to Wi-Fi access when Wi-Fi becomes available.

The switching between access networks is currently managed by the user device only or supported by protocols given by the mobile service operators. The protocols are kept in an access network discovery and selection function (ANDSF), in the core network and communicated to the user's device. The user device is responsible for implementing the protocol, i.e., selection of the appropriate air interface.

More efforts are ongoing by 3GPP to further supplement in an access network discovery and selection function (ANDSF) based techniques with radio access network (RAN) information [19].

Figure 3.7 shows the evolution of these techniques, from technique selection based on User Equipment (UE) to selection enhance by radio access network (RAN) signaling, and then to the simultaneous use of both interfaces by an application in order to further reduces service delays and consistently Maintains user [20]. Many other studies focused on how to integrate Wi-Fi with LTE Technology; also, Wi-Fi uplink spectral efficiency and coverage limitations were discussed, as Wi-Fi cost-effectively addresses the tremendous demand for data from wireless devices by leveraging unlicensed spectrum. Different options for the integration were offered as a result of expanding the LTE with the presence of multi-radio devices with different levels of implementation complexity and the corresponding trade-off in benefits. The integration enables an effective traffic offload in the LTE network and serves more users with higher throughput demands through the Wi-Fi networks; meanwhile, the frequency spectrum, bandwidth, security, and various factors pose a trade-off issue for the integration. The mutual compliments between the software-defined radio (SDR) and software-defined network (SDN) are playing an essential role in having easy integration between LTE and other networks (IP network & Wi-Fi). The performance, power saving, security, and optimization problems derived from the interactions of the controller and two-layer policy should be explored with more considerable effort in the future, as shown in figure 2.8 [21].

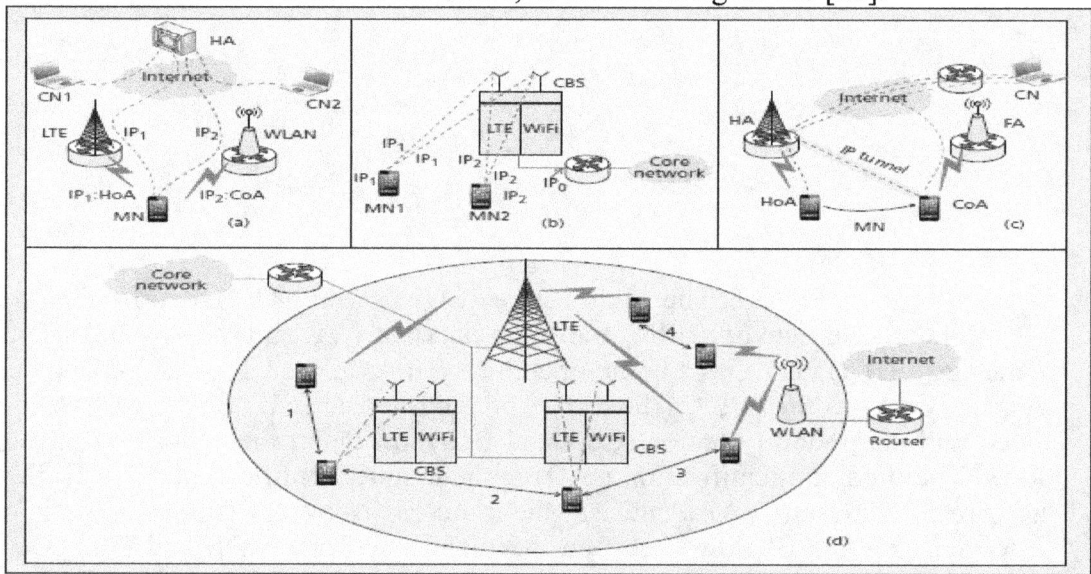

Fig. 3.7. The Heterogeneous Network Scenarios

Many other studies focused on how to integrate Wi-Fi with LTE Technology; also, Wi-Fi uplink spectral efficiency and coverage limitations were discussed, as Wi-Fi cost-effectively addresses the tremendous demand for data from wireless devices by leveraging unlicensed spectrum.

Different options for the integration were offered as a result of expanding the LTE with the presence of multi-radio devices with different levels of implementation complexity and the corresponding trade-off in benefits.

The integration enables an effective traffic offload in the LTE network and serves more users with higher throughput demands through the Wi-Fi networks; meanwhile, the frequency spectrum, bandwidth, security, and various factors pose a trade-off issue for the integration.

The mutual compliments between the software-defined radio (SDR) and software-defined network (SDN) are playing an essential role in having easy integration between LTE and other networks (IP network & Wi-Fi). The performance, power saving, security, and optimization problems derived from the interactions of the controller and two-layer policy should be explored with more considerable effort in the future, as shown in figure 2.8 [21].

Fig. 3.7. The Hybrid Architecture of SDN and SDR Configurations

The integrated model consists of 2 parts:
- A centralized controller is installed for all IP packets which are carried by software-defined network SDN, Wi-Fi SDR, and mobile network
- Small-cells access points can be used instead of macrocells to reduce the cost of data rates, and This can be done by replacing LTE macrocells by large numbers of small access points.

The architecture consists of three main components:
- Access nodes, an evolved NodeB (eNodeB) for 4G; NodeB with radio network controller (RNC) for 3G (defined by the 3G partnership project, 3GPP, standards) and access point for Wi-Fi (defined by the IEEE 802.11 standards);
- Edge switches (ESs) with an edge control plane (ECP), which serve as both ingress and egress elements for network service provisioning;
- The 4G forwarding control plane and infrastructure.

Applying the fabric approach is the crucial property of SDN design, by the separation of ECP and ESs from the 4G network. Through this separation, most of the intelligence to ECP is limited, keeping the 4G network unchanged, and this means that ECP is in charge of providing productive network services, while the 4G network only performs packet transmission through the 4G infrastructure. Additionally, this separation allows an independent evolution of the ECP and 4G network, focusing on the specifics of each role. Under this separation, incoming service requests are processed, following three phases in figure 3.8 [22].

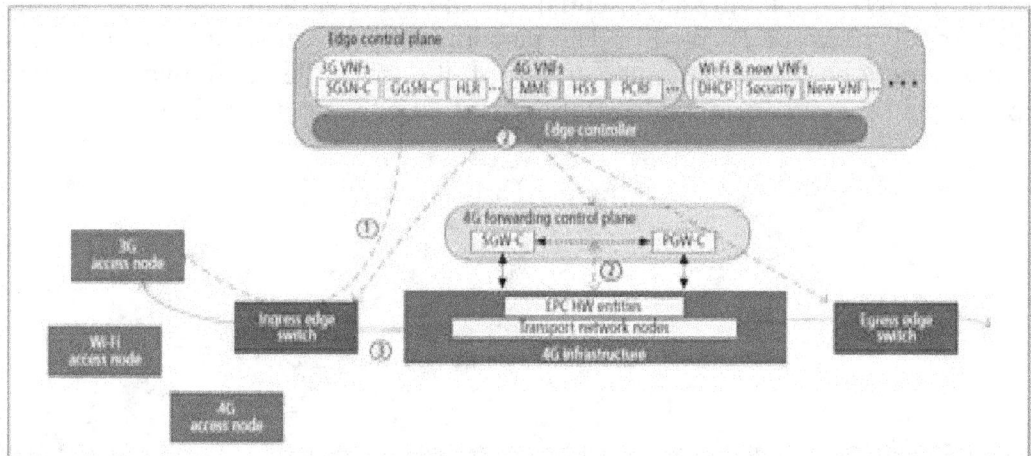

Fig.3.8. The SDN for IP Core Network and Mobile Flatter Architecture

In the first stage, certain network functions are performed with the specified virtual network functions (VNFs) in the ECP. After that, the ECP interacts with the 4G forwarding control plane and ESs to set the end-to-end data path. Last, data is delivered.

Five types of interfaces are used to design the architecture:

Application-operator, which is used to determines how services can be provisioned by the network Operator.

Operator-operator, this interface decides how different domains inform each other with their requirements

Operator-network, host-network, and packet switch interfaces are respectively mapped to the northbound, horizontal, and southbound interfaces, as defined in the general SDN framework, as shown in figure 3.9 [23].

It is assumed that the proposed integrated model of software-defined network SDN, Wi-Fi SDR, and mobile network already deployed access nodes as they are; this is because:

- There are much more complex and challenging efforts to split the control functionality from the access nodes than for the other core nodes.
- Because of cost constraints, Modifying all deployed access nodes is not practical, as shown in figure 3.10 [22].

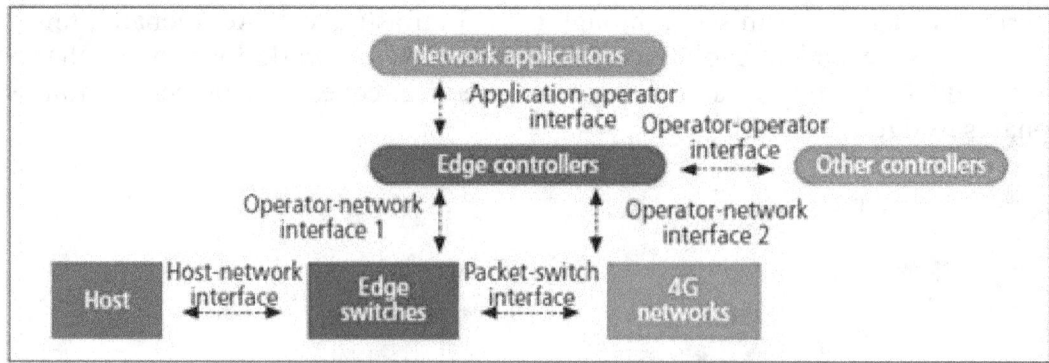

Fig. 3.9. The SDN-CM Interfaces

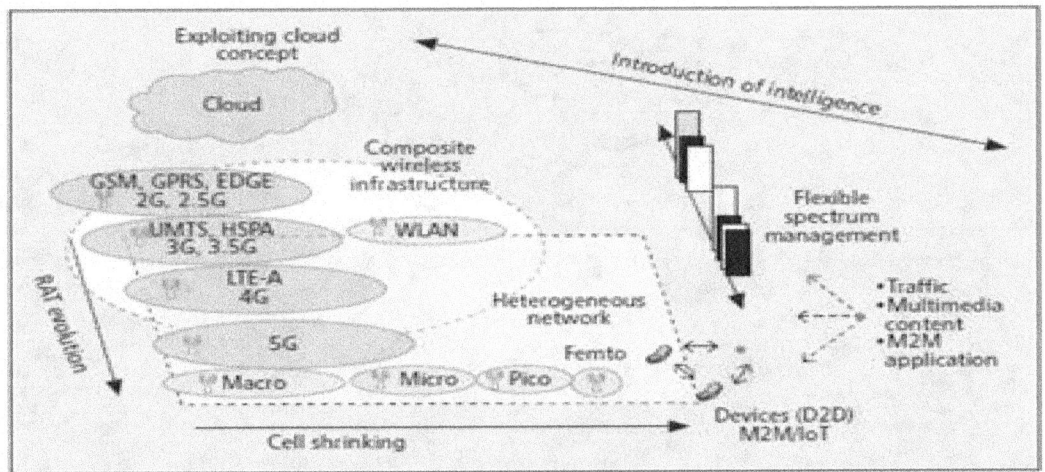

Fig. 3.10. The Evolution of the Mobile Network and the Key Technologies

The integrated model of software-defined network SDN, Wi-Fi SDR, and mobile network to the wireless parts is done by using LTE small cells, which provide higher data rates to users than the LTE macrocells. The channel quality at the receiver will be better due to the short distance between the user and the small cell. The Wi-Fi cell provides additional capacity, but it also has a limited coverage range. At the same time, the integration between LTE and Wi-Fi and the macrocells is shrinking [22].

3.1.3. Integration with IP Multimedia Subsystem (IMS)

The IP Multimedia Subsystem (IMS) directs the communication technologies to the next level by providing enriched communication techniques, as it offers robust multimedia services through roaming boundaries and by using diverse access technologies. The IP Multimedia Subsystem (IMS) model is designed for supporting multimedia services and provides communication over different access technologies. The IP Multimedia Subsystem technique is a collection of certain functions connected by standardized interfaces. The architecture of IMS consists of 3-layers, as shown in figure 3.11 [24]:

- The transport layer, which contains all the entities for the supported access networks.
- The control layer where the core IMS network has resided.
- The service layer which includes the application servers hosting the IMS services.

As the IMS networks use the Session Initiation Protocol (SIP) (a protocol for signaling and controlling multimedia communication sessions) for session establishment, management and transformation, the ability to combine and identify a set of IP-based services in any case will be exist during a single communication session and can mux and de-mux any other services. The traditional model mixes an LTE core network, a WiMAX network and a WLAN network and connects with the LTE core through identified functional structures and an IMS in a duty of session's control. Thus, the clients can access the LTE Circuit-Switched (CS) services through the WiMAX and WLAN networks, since they are authenticated in the AAA (Authentication, Authorization, and Accounting) server and recorded in the IMS core [25].

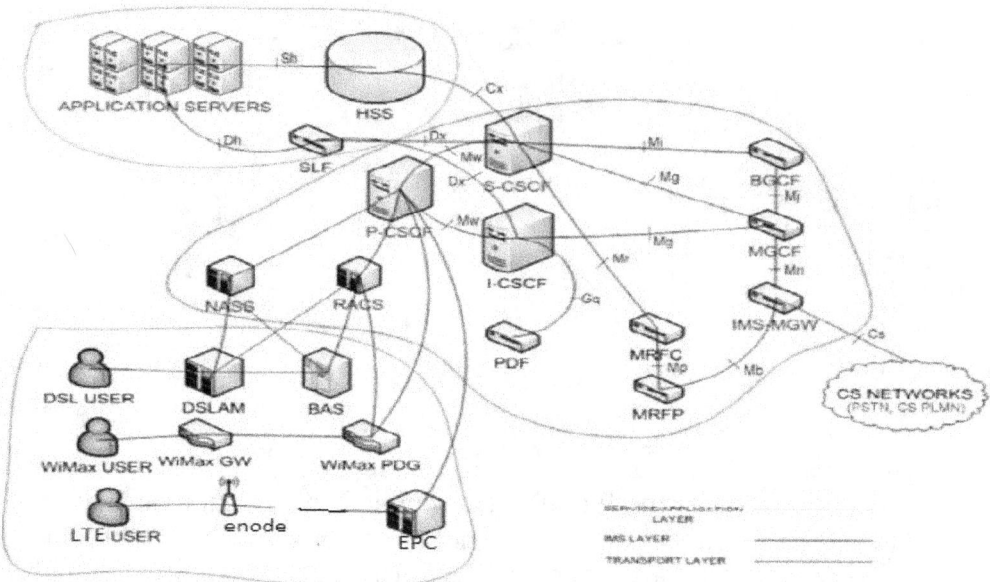

Fig. 3.11. The IMS 3-Layer Architecture.

The IP Multimedia Subsystem (IMS) is a universal, access-independent, and standard-based IP communication and service control structure that enables different types of multimedia services to end-clients through using standard Internet-based protocols. The IMS has been adopted by the 3GPP as the base of the ITU work in Next Generation Networks (NGNs), which is the body that took the Session Initiation Protocol (SIP) as the control protocol for multimedia communication and has built the finite structure for SIP-based IP multimedia service machinery.

Since the IP Multimedia Subsystem uses the SIP for session establishment, management, and transformation, it offers functions, such as authentication, addressing, routing capability negotiation, service invocation, provisioning, charging, and session establishment. In addition to the SIP, many other protocols play essential roles in the IMS as Diameter and HTTP. [26].

Other studies focused on the integration with IP multimedia subsystem (IMS) infrastructure, services, and applications. They proposed the next generation of the heterogeneous network by the merged architecture with UMTS-WiMAX-WLAN tight coupled architecture along with IMS, and the approach is driven by upgrading the network from UMTS to LTE in the framework of the already established heterogeneous network (UMTS-WiMAX- WLAN) tight coupled architecture. The proposed architecture gives high performance during data transfer and less time during IMS registration and IMS session establishment. The integration between LTE Network and IMS system is extending IMS beyond 3G, and this offers the users various networks with high-quality IP-based multimedia service, and it also has the effect of improving the performance of the heterogeneous network. In the future, it is necessary to consider solving IMS-based proposed architecture open issues concerning session control, authorization, charging, and personal mobility, as shown in figure 3.12 [27].

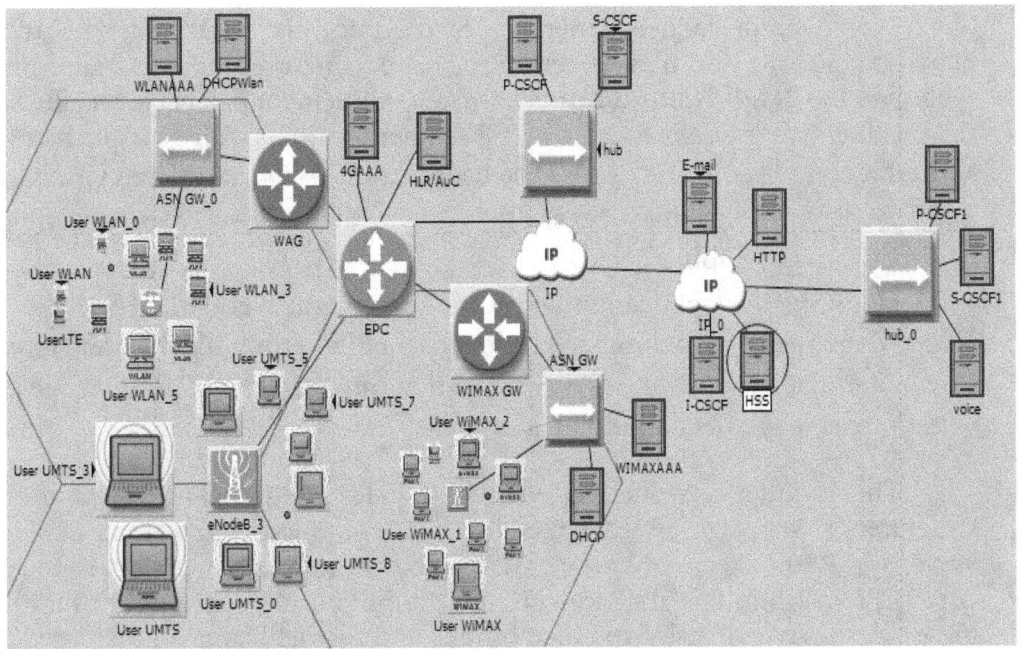

Fig. 3.12. The LTE-WiMAX-WLAN Tight Coupled Interworking Architecture

3.2. Security in backhaul section of LTE Network

Because the 4G is an IP-based network, there are remaining many extensive studies on how to secure 4G networks. The 4G networks find an open environment where different network operators and service providers use the same core resources through open interfaces, open hardware, and software platforms for the end-user's equipment. This open environment of the 4G demonstrates much more security difficulties as opposed to the traditional structure of the closed environment (e.g., public switched telephone network, PSTNs), which has the advantage of protection against security threats. There are many studies on how to find and design new security solutions for the 4G network to guarantee a high level of security for the infrastructures and services to be fully protected against all the different types of threats, as well as offer to the end-users secured accesses to the requested services.

In the first generation (1G) mobile system, there was not much value for security; there was no over-the-air encryption in the system. In the mobile phones could be efficiently trapping the serial number and the eavesdropping the conversation. The 2G, represent by Global System for Mobile (GSM), depends on Authentication and Key Agreement (AKA), which called GSM AKA for encryption and authentication. It uses a challenge answer technique, where, the user confirms its identity by offering a response to a time-variant challenge raised by the network. However; its security is weak in that:

- Only unidirectional authentication is used
- The serving network cannot be authenticated by the user
- The authentication information and cipher keys can be reused, and the authentication data (triplets) can also be reused

The Authentication and Key Agreement (AKA) was developed in 3GPP by allowing the mutual authentication and agreement on an integrated key between the mobile terminal and the serving network, and vitality assurance of agreed cipher key and integrity key.

Moreover, a sequence number is used for the vitality where two counters (one for network and one for mobile terminal) have concurred for sequence number confirmation. Although the 3GPP Authentication and Key Agreement has been accepted as reliable and used, there still existing weaknesses in 3GPP Authentication and Key Agreement as:
- Redirecting user traffic using false base station (BS) and mobile terminals.
- The counter value may be set to a high value by scamper; the mobile terminal's lifetime may be abbreviated.
- In the 3GPP Authentication and Key Agreement, Networks keep a counter database and dynamically synchronized for every mobile terminal (MT), and any mistake in the counter database with resynchronization processes which is requested by mobile terminals may affect all mobile terminals.

The mobile networks have been transferred to all-IP based transport network structure. It supports IP based end-to-end communication from evolved Node base stations in Radio Access Network (RAN) to core network elements in the evolved packet core (EPC). While this flat structure has simplified the operation of mobile networks, at the time, the overall risk of vulnerabilities and threats has been increased. Therefore, the challenge is to find a full and effective security model which can offer isolation for any fault as well as to secure the internetworking of legacy and non-3GPP networks. The 3GPP has specified the LTE security mechanism that provides the security features, techniques for each section of the evolved packet core (EPS). Figure 3.13 illustrates the 3GPP security architecture [28].

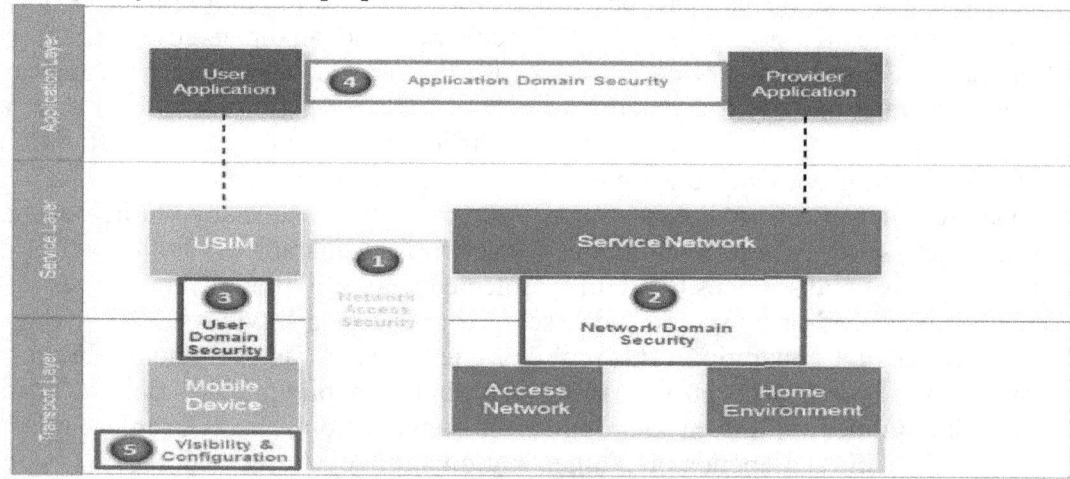

Fig. 3.13. The 3GPP Security Architecture

The 3GPP security construction has been organized into five classes as follows:
- Network access security: validates that mobile users have secure access to network services and mobile network is protected against any attacks through the radio access link.
- Network region security: verifies that mobile backhaul nodes transfer signaling data and user data vulnerable network and defends it against any attacks through a wire-line link.
- User domain security: protects access to mobile stations

- Application domain security: allows applications on a user and network side to exchange data securely.
- Visibility and configure of security: allow customers to have information about enabled protection features and provision of services.

The LTE security form consists of various security systems at different levels, as shown in figure 3.14 [28]. Besides, LTE introduces new security techniques, such as crucial derivation mechanism during mobility (Key of Access Security Management Entity, "KASME"), protection of radio interfaces frames, user session ciphering, and Radio Resource Control "RRC" with radio signaling integrity control and ciphering). LTE introduces a new protocol called X2 to exchange the user and signaling data at the air interface of the 3rd Generation Partnership Project (3GPP) of LTE, which called the Evolved Universal Terrestrial Radio Access (EUTRA). The General packet Radio Service Tunneling Protocol (GTP) and IP security tunnels are used to secure the X2 interface. The evolved packet core (EPS) also adapts 2G/3G security techniques to obtain a suitable security structure by establishing confidentiality and integrity algorithms in the EPS protocol stack. For instance, access stratum (AS) signaling integrity and encryption protect the Radio Resource Control (RRC) protocol [29].In the LTE security structure, the Mobility Management Entity (MME) identifies user equipment (UEs) by using the authentication data from the home network and triggering the Authentication and Key Agreement (AKA) protocol in the user equipment (UE). This operation will share the Key Access Security Management Entries (KASME) key. Further keys can be provided for confidentiality and integrity security at the Non-Access Stratum (NAS) level. These keys are used to secure signaling and user data in many evolved packet core (EPC) interfaces [30].

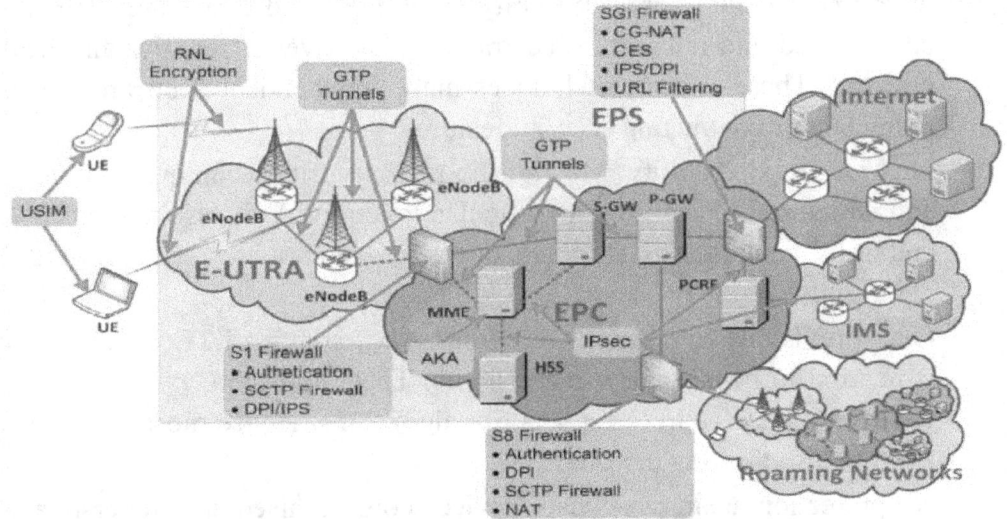

Fig.3.15. The LTE Security Architecture

The LTE Security is structured on the previous 2G/3G security techniques with some develops that contain better security algorithms, longer keys, extended vital structure, and the introduction of new features to address backhaul and relay node security. Although, it inherits several of the defects basically due to the complexity and fragmentation of the security structure, the reliance on distributed and uncoordinated security techniques, and its decentralized control. Other defects are as follows [31]:

- Open Mobile Networks: before LTE, the mobile networks began as a closed system with identified security controls built-in to the communication protocols. Extensive efforts were needed to break these systems.
- On the other hand, LTE and new mobile structures were implemented to be more flexible and openness to the outside networks such as partner roaming networks, IP multimedia subsystem, and the Internet. All voice and data services are now availed through the IP protocol. This openness has provided agility to the overall network, and on the other hand, it has risky to new threats such as IP spoofing, network worms, and quality of service attacks.
- Place in Network (PIN) Based Security Model: LTE security structure is a point-based algorithm. Such, Deep Packet Inspection (DPI), Customer Edge Switching (CES), Network Address Translation (NAT), Intruder Detection Systems (IDS), and firewalls are implemented only at the Internet, roaming borders and mobile access. All of these techniques offer to set up security functions for network points. Thus, security is intensively concentrated on externally originated attacks, while the little concentrate is implemented to address the internally originated attacks.
- Isolated Security Models: A various set of non-coordinating security techniques are implemented in different areas of the mobile network. However, the majority of the security functions are done on an isolated island with an identified task that is designated to that security component. As an example, the S1 firewall is defined as Evolved Universal Terrestrial Radio Access (3GPP) (E-UTRAN) to the evolved packet core (EPC) border. Security is implemented in a fragmented way that offers a complicated and uncoordinated frame of the overall security and health of the LTE network. These security techniques make independent decisions and may implement redundant or contracting security features.
- Lack of interoperability: Most of the LTE security techniques are vendor dependence and structured to work on a particular function. As a result, such models are closed, and it is complicated to obtain a "mixed and matched" use of different security algorithms.
- Over-provisioned security resources: nowadays, LTE security techniques are structured to handle busy traffic hours. They stop the service inconvenience by over qualifying the resources for security algorithms. Hence most of the security resources are little use for long periods.
- No protection against backhaul devices compromised and impersonated attacks: LTE networks use small cell-based stations such as mobile femtocells, which are implemented in clients' buildings. These small base stations are not physically protected in the same manner as a traditional base station. The majorly of these microcell base stations are not managed by the operators, and they are highly vulnerable to unauthorized tampering.

However, these shortages of the security system do not prevent the scalability, flexibility, and adaptability, but also increase the deployments and operational cost of the legacy LTE security techniques. However, future mobile networks demand advanced, intelligent, and collaborative security systems to recover the above defects. The conventional strategy of the network security has concentrated on securing the network boundaries to block outside threats from reaching the network resources which is not adequate because the attackers attempt to detect protection vulnerabilities in networking protocols, administering systems or applications, and employ these vulnerabilities to scatter malware that may avoid security measures at the edges. The Possible threats to 4G networks cover IP address spoofing, user ID theft, theft of service (ToS), denial-of-service (DoS), and intrusion attacks. Between them, network administrators are concerned about ToS and DoS attacks because they will harm their wealth, reliability, and service availability. The security threats are further classified, according to X.805 as follows [25]:

- Removal of information and other resources,
- Corruption or adjustment of information,
- Theft, removal or loss of information or other resources,
- Disclosure of information,
- Interruption of services.

Besides this general categorization, protocol-specific attacks must be recognized. For example, SIP-targeted attacks include:

- Malformed message attacks,
- Buffer overflow attacks,
- Denial-of-Service (DoS) attacks,
- RTP session hijacking,
- Injection of unauthentic RTP
- Reuse of Compromised SIP credentials,
- Bogus SIP network elements.

It is almost painful to make a 100% secure system because new threats and vulnerabilities will remain to take place. Also, there exist many stakeholders, including at least network operators, service providers, and users, having their own, sometimes respectively conflicting, business, driving to various security conditions. Therefore, the 4G security structure must be manageable enough to adjust itself to coming threats and vulnerabilities as well as diversifying security requirements [25, 28].

Several types of research discussed the security tools of the 4G Network and all exposed, as the 3GPP general packet radio service (GPRS) core network supports 2G GSM and 3G mobile networks to have IP interconnectivity with conventional networks such as the Internet, the Security algorithms expert model (SAE) is the development of the GPRS infrastructure, which enables LTE to become practical within the 4G technology. The proposed architecture reduces the number of nodes and manages better share the processing load, minimizing latency in the network, as shown in figure 3.16 [26].

Related to the conventional scheme, which includes more item nodes (four in entire), for simplicity, the recommended structure has only two network elements enhanced NodeBs and evolved packet core (EPC) that perform all-IP mobile core networks for 3GPP LTE.

At the same time, the control plane is separated through a well-defined open signaling interface (S11) among the MME node and gateway.

The mobility management entity (MME) unit acts the essential function in the control methods of the LTE/SAE, and it is granted as a part of the endpoint for the signaling (control) plane interfaces of SAE, as well as SAE's ciphering/integrity protection tools, plus it is accountable for two steps in the security system of LTE which are the user authentication and the optimal selection of the Serving Gateway (SGW) for user equipment (UE) at the initial attaching phase. The Serving Gateway (SGW) acts as a mobility mooring for both inter and extra handover connectivity across evolved Node B (eNodeB) and other 3GPP technologies. It transmits and routes packets while the PGW produces the connectivity for the mobile user to outside packet data networks. It is deserving seeing that the home location register (HLR) of the original (GSM; 2G) and universal mobile telecommunication system (UMTS; 3G) structure is enlarged to the home subscriber server (HSS). The static subscriber information is maintained there and integrated with the authentication center (AUC). The AUC also generates temporary authentication and security data that can be employed for subscriber authentication and user traffic encryption. For the stable and compatible operation of the LTE/SAE network structure, it is needed for the earlier described objects to relate each other with source or endpoints. We can distinguish them as continued or dashed unique lines between the network entities. Links, with a continuous line on them, are used as user traffic transmission connections; whereas dashed links are used for signaling targets [26].

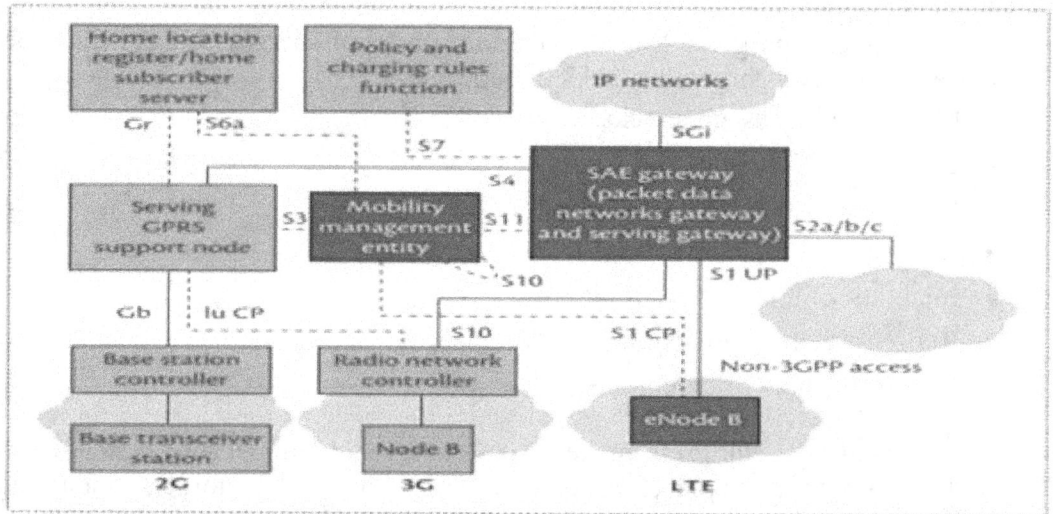

Fig. 3.16. The SAE Evolved Packet System's (EPS) Architecture

The 3GPP generates tags for each reference point, such as:

- S6a point interfaces among MME and HSS and utilized to change data which is associated with the position of the mobile station and the control status of the subscriber
- SGi point might be outside or intra-operator packet data network.
- S10 interface between multiple MMEs supports MME transmission of user information and MME relocation.

To contribute a mutual authentication system between the UE and MME through the evolved UMTS terrestrial radio access network, the 4G wireless access network (LTE/SAE) reuses UMTS authentication and key agreement (UMTS-AKA), this authentication protocol, called evolved packet system authentication and key agreement (EPS-AKA), is utilized to produce all the ciphering and integrity keys needed to guarantee encryption and integrity security. All the keys are determined via the key derivation function. Although there are several known security vulnerabilities on the underlying AKA protocol, the UE/universal subscriber identity module (UE/USIM), the card embedded into subscribers' mobile phones used for their identification and authentication, always implements an AKA algorithm to generate the mutual authentication with mobile core networks. Although the authentication protocols might be different, they all perform the AKA algorithm on UE/USIM. Many kinds of research are still carried out by The security algorithms group of experts (SAGE) of the European Telecommunications Standards Institute (ETSI) to set up the 4G standard of LTE/SAE, and they listed the basic 3G algorithms for encryption and integrity protection for the LTE network standards as follows [28]:

- **The Kasumi Algorithm:**

It recognized as the primary algorithm for the LTE ciphering standard, which applies a robust encryption algorithm via 128 bits key. It applies the fundamentals of the 3G conditions and gives excellent protection to the most traditional block cipher attack methods.
Additionally, it employs two mapping functions called S-boxes to produce the cipher text. It was mainly designed as a building block for the UMTS encryption algorithms (UEA1) and integrity algorithms (UIA1), the most vulnerable points of this algorithm are:

- It has lake adequate protection against the new methods of the algebraic attacks
- It needs a tradeoff among the performance and the hardware implementation complexity.

- **SNOW 3G:**

It was invented a 3G network as a second cryptographic solution in response to the presence of more modern methods of attacks such as algebraic attacks, which were not included by Kasumi-based algorithms' security. The original SNOW algorithm was adapted by some changes to meet the requirements the 3G demands, and to defend against the newly invented algebraic attacks the weakest point of this algorithm is:

- It needs more complex computational methods in terms of the hardware area space.

- **The Mileage Algorithm:**

This algorithm met the demands of the 3G network and performed in the LTE network as the 3G security algorithm number 3 that uses a key size are 128 bits. It gives as secure implementations versus the side attacks using standard Advanced Encryption techniques as a core function, the weakest point of this algorithm are:

- It needs Joint actions of different universal subscriber identity module implementations

- **The ZUC Algorithm:**

It is recognized as the most complicated algorithm from the computational point of view and the hardware area space. It offers superior protection toward the new forms of algebraic attacks. It bypasses the same design principles, Mileage, and Kasumi algorithms. It provides secure implementation and protection against side-channel attacks via Advanced Encryption Standard as a core function using secure encryption via 128 bits key. Also, it was constructed on well-known ciphering algorithms systems, the weakest point of this algorithm is:

- It needs more investigations to get farther assurance.

Many studies are proposed to use two types of the previous algorithms in the LTE network (i.e., the Kasumi algorithm and SNOW for the threats of 4G) to overcome the possible threats of the 4G network.

3.3. Capacity in the backhaul section of LTE Network

The 4G technology raises the demands for the capacity of the backhaul in the LTE network to gigabits per second, which enhances network utilization and reduces the operations costs, and applying the carrier Ethernet transport construction at a lower cost of ownership than the legacy time-division multiplexing (TDM) transport infrastructure. The essential strengths behind this are the mobile broadband services and the narrowband circuit-switched data networking that confirmed the general packet radio service (GPRS) in the second-generation (2G) systems, and the high-speed packet access (HSPA) in the third-generation (3G) systems. The enormous expansion of the broadband data services caused a significant transfer in the construction of mobile traffic from voice-dominated circuit-switched traffic to packet-switched data traffic. The deployment of LTE, which can present a hypothetical peak downlink data rate of 330 Mb/s (i.e., peak rate of LTE radio base stations [RBSs] with 4 × 4 multiple-input multiple-output [MIMO] antenna configuration), will further improve the relative variation in the needed capacity between packet data traffic and TDM traffic in the backhaul as shown in figure 3.17.

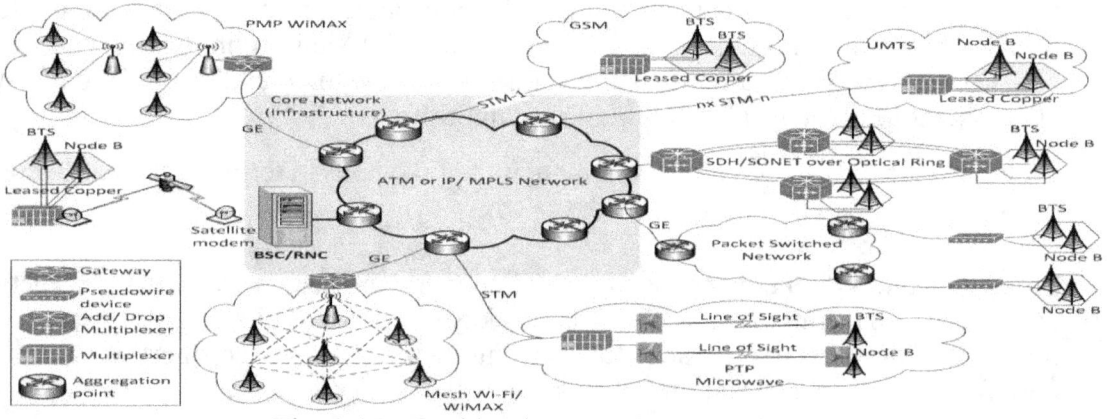

Fig. 3.17. Backhaul Network Technologies

The conventional backhaul technologies contain:

- Copper and Optical Fiber Backhaul Networks:

The Copper cables are the essential backhaul medium between Base Transceiver Stations (BTSs) and Base Station Controller (BSC) in the 2G and 3G mobile systems. Time-division multiplexing techniques that are executed in Plesiochronous Digital Hierarchy (PDH) technology were the universal tool that allows multiplexing of many voice subscribers from base stations and transferring them to the A base station controller (BSC) in different time slots. There are two models of plesiochronous digital hierarchies model, which mostly differ in the delivered bit rates [1]: The T-carriers leased lines (T1, T2, T4) and E-carriers leased lines (E1,..., E5). The T1 circuits work on 1.544 Mbit/s, while E1 circuits work on 2.048 Mbit/s. The T-carriers are found in North America and Japan, while E-carriers are found in Europe and the rest of the world. T1/E1 circuits can be communicated as point-to-point systems or over plesiochronous digital hierarchies (PDH) transmission systems. The T1 frame consists of 24-time slots; each can support a 64 kbit/s voice channel. Considering a one-time slot is defined for signaling; however, without any voice compression, a T1 line can carry 23 voice subscribers. For E1, there are 32-time slots where 30-time slots are used for voice subscribers, and the other two are used for frame alignment and signaling. For better bandwidth efficiency, voice compression mechanisms such as G.729 are provided to compress 64 kbit/s encoded voice words, resulting in utilization gain of four times than the traditional model, as shown in figure 3.18. In several cases, the circuits of T1/E1 copper lines from multiple sites are multiplexed from lower rate T1/E1 connections into higher rate connections such as Synchronous Transport Module level-1 (STM-1) with bit rate 155.52 Mbit/s, STM-4 (622 Mbit/s) and STM-16 (2.5 Gbit/s). The Synchronous Transport Module (STM) standards are used in North America as Synchronous Optical Networking (SONET) and in Europe and the rest of the world as Synchronous Digital Hierarchy (SDH). Time Division Multiplexing (TDM) in the backhauling of a mobile network can communicate synchronization information throughout the network [31, 32].

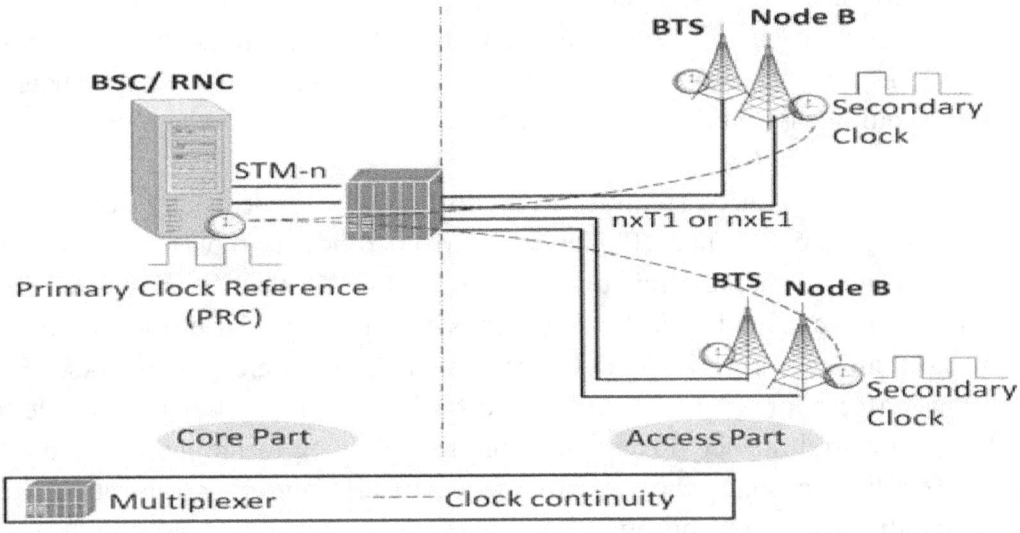

Fig. 3.18. The TDM Backhaul Network of Point-to-Point Leased Lines

The add/drop multiplexer is an essential node of the transmission network, which used to multiplex many lower data rates data frames into a single frame with a high data rate frame. While at the same time, add or drop other low bandwidth signals by dropping them from the frame and forward them to other network routes. The optical fiber rings represented in SDH technology can be used to justify the cellular sites with the requested data rates in case of the cost conditions are available, where it was proposed to replace T1/E1 circuits at the mobile sites based on a backhaul cost model. The cost model includes the distance between the cell site and SDH multiplexer, the number of T1/E1 circuits per site, and the number of base stations to be connected through the optical SDH ring. The copper lines support enough bandwidths for voice traffic in the 2G mobile system. However, to provide the data rate needed for 3G traffic and after, many parallel circuits are needed, which results in a cost-increasing linearly with the provided bandwidths. For large bandwidths, copper lines become very costly and are not a suitable offer for the backhaul upgrade of next-generation systems such as 5G [33, 34].

Microwave and Satellite Wireless Backhaul:

Microwave radio is the other technology that is used in the backhaul network design, as while it contributes by 6% of the total transmission bandwidths in the United States. Microwave radio circuits are a substitution for wired backhaul circuits principally in the aggressive geographically areas where wired connections are impossible to be implemented. Microwave models can be executed in different frequency bands containing licensed (6 GHz to 38 GHz) and unlicensed (2.4 GHz and 5.8 GHz) bands. Applying the unlicensed bands can reduce capital expense (CAPEX) but increases radio interference problems. According to the used frequency spectrum, the bandwidth capacity and distance coverage will be affected as using higher the frequency, the bandwidth capacity increases, and the coverage range will be shorter. In all situations, the existence of a line of sight (LOS) between cell sites and aggregation points is the need, and hence microwave is limited to the short-distance transmission when used in metropolitan environments.

Moreover, in the open environments, when the line of sight (LOS) exists, microwave links can be easily implemented to cover the long distances transmission. A compared copper system, the microwave system needs a higher capital cost due to equipment expenses and spectrum licensing costs. On the other hand, the microwave system provides lower Operational Expenditure (OPEX) over a long time [35].

Many advanced solutions to the difficulty of the capacity between the access port and the core port in the LTE network explained how to replace the low capacity links of copper leased lines and microwave links with the Optical fiber system. The Existing and emerging optical fiber technology finds two principal insertion points that enable the maximum performance of 4G integrated fiber-wireless networks. Existing optical transport networks (OTN) can be used for baseband data transport in the backhaul , and the optical millimeter-wave generation are emerging technologies for the front haul that can utilize existing fiber to the home (FTTH) and passive optical network (PON) capabilities for 100 Gb/s aggregate, multi-service, point to multipoint radio transmission over optical fibers [28].

The best implementation of this model is to use the time wavelength division multiplexed passive optical network (WDM-PON). The architecture of the TWDM-PON was selected by ITU–T as the next-generation optical access network technology. The WDM-PON technology designed to support the access system by multiple 10-gigabit passive optical networks (XG-PONs) via stacking them in multiple pairs of wavelengths. Figure 3.19 shows an example of the TWDM-PON system with four stacked XG-PONs. Each XG-PON operates in a pair of downstream and upstream wavelengths [35].

The rate of an XG-PON is 10 Gb/s in the downstream, and 2.5 Gb/s in the upstream, and the system can reach the rate of 40 Gb/s in the downstream and 10 Gb/s in the upstream. The WDM-PON contains tunable transmitters and receivers; the tunable transmitter can tune to any of the four upstream wavelengths, and the receiver is tunable to any one of the four downstream wavelengths. WDM-PON has already proved itself as a promising candidate for future optical broadband access systems and next-generation mobile backhaul networks. It could meet the requirements for achieving a high data rate, connecting a large number of subscribers, and reducing CAPEX with the drastic growth of beyond 4G mobile data traffic and ever-increasing demands on high-speed broadband access services [31].

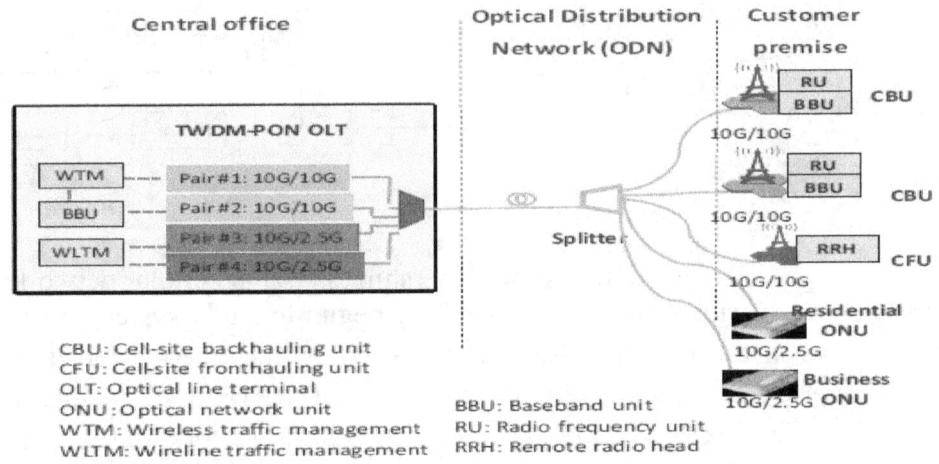

Fig. 3.19. The Wavelength division multiplexed passive optical network

3.4. Energy consumption in LTE 4G Network

The wireless access network mobile systems are considered the most significant part of the network consuming energy. Throughout the period from 2007 to 2012, there was an expansion of 10% in using energy per year. The energy consumptions by wireless access networks will be raised in the next few years, with the extreme increases of the mobile networks and the need to extend in the conventional technologies of the 4G and 5G mobile systems. Furthermore, green networking will be essential for the deployment of prospective wireless access networks [36, 37].

Traditionally, in mobile methods, actually individual researches have been conducted to improve the utilization of the power consumption in the user devices, to increase its battery life.

Now there are several studies in mobile communications that extended its attention to power consumption at the base station (BS) side. The user equipment utilizes about 10% of the energy consumed by base stations, and these base stations take about 60% and 80% of the total mobile network energy consumptions [38, 39]. By finding lower energy consumption in the access nodes, such as small cells and relays, mobile operators move to provide the expanding requirements of the data traffic and reducing the power consumption in the base stations by trying to reducing the range between the end-users and the serving access point (AP). To evaluate the effect of different system elements on wireless networks' energy performance, and to select potential best solutions, it is necessary to recognize professional models that can contribute proper consideration of energy consumption. It is also necessary to consider the reality of the overall total cost of ownership (TCO) of the mobile network, as shown in figure 3.20 [40].

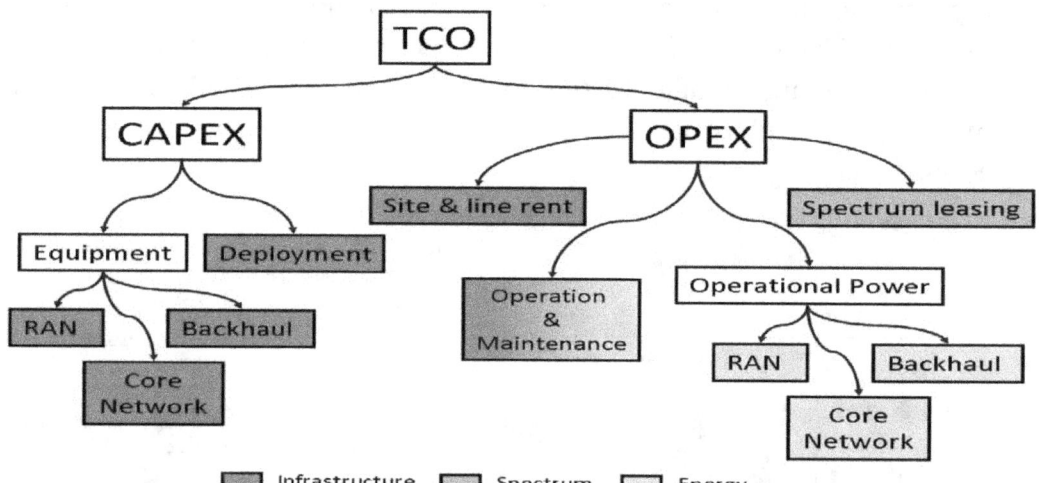

Fig. 3.20 The total cost of ownership (TCO) in cellular networks

Consequently, various studies have recently begun to study systems that can describe the associations between the principal sources of expenses and energy consumption in wireless networks. Though, the enormous implementations of RRHs will require excellent energy consumption if every RRHs are initiated even with low network traffic. Also, if each RRH assists all users' equipment, significant energy will be used on the front-haul links. As a consequence, how to initiate the RRHs and select the RRHs for accepting user equipment to decrease the total network power consumption (NPC) is a significant problem [41].

The proposed solution to reduce the operational cost of the energy consumption in the LTE network is to transform the network from macro-cells to small-cell based cloud radio access networks (C-RAN). The solution offers the best access control, interference control, supervision of the network along with higher throughputs and capacity. The solution also offers a cost-effective solution and scalable in a challenging dense civilized environment, as shown in, shown in figure 3.21.

Another form of the proposed solution is the Heterogeneous cloud radio access networks (H-CRANs), which is proposed as an optimized potential solution to mitigate inter-tier interference and enhance mutual processing gains in HetNets through m with cloud computing. The motive of Heterogeneous cloud radio access networks (H-CRANs) is to improve the capabilities of VPNs with enormous multiple antenna techniques and simplify LPNs through gathering to a signal processing cloud with high-speed optical fibers. As a result of this technique, the energy consumption of the wireless infrastructure is decreased, and the operating expenses are reduced. Figure 3.22 shows the improvement milestone from the traditional 1G to the H-CRAN according to 5G technology. In the traditional first, second, and third generations (1G, 2G, 3G) mobile systems, mutual processing is not required because inter-cell interference can be eliminated by exercising the static frequency planning or code-division multiple access (CDMA) mechanism [41].

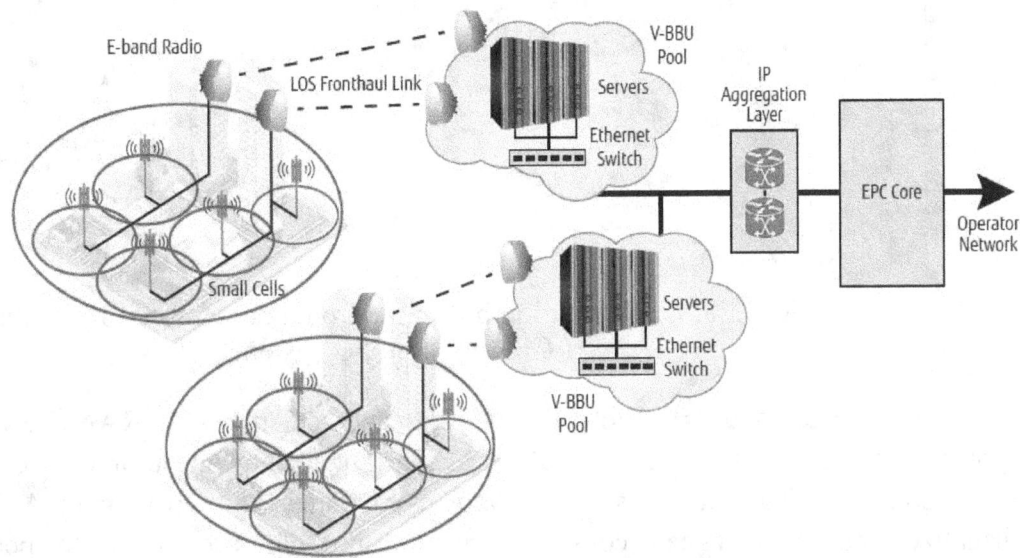

Fig. 3.21. A scalable and cost-effective CRAN Architecture in a challenging dense urban environment

However, the inter-cell interference is severe in the orthogonal frequency-division multiplexing (OFDM)-based 4G systems, because of the spectrum utilization of the adjacent cells, especially when a HetNet is diffuse. Therefore, inter-cell or inter-tier cooperative processing through CoMP is severe in the 4G network. Cooperative communication mechanisms have developed from two-dimensional CoMP to three-dimensional large-scale cooperative processing and networking with cloud computing to manage more and more critical interference and enhance SE and EE performances. Therefore, for the Heterogeneous cloud radio access networks (H-CRANs) based 5G system, cloud computing-based cooperative processing, and networking mechanisms are offered to look over the mentioned challenges of 4G systems and in a switch to the performance demands of 5G systems [40, 41].

Although the HetNets are the best solutions to provide smooth coverage and high capacity in 4G systems, there are still two critical challenges to prevent their commercial improvement:
- The SE performance should be improved because the intra / inter-cell CoMPs need a large amount of signaling in backhaul links to reduce the interference between the

HPNs and LPNs, which often makes the capacity of backhaul links is over the limited band.
- Ultra-dense LPNs can enhance the capacity at the expense of consuming too much energy, which results in low EE performance for providing high EE together with gigabit data rates across software-defined wireless communication networks. In this case, the virtualization of communication hardware and software elements place stress on communication networks and protocols.

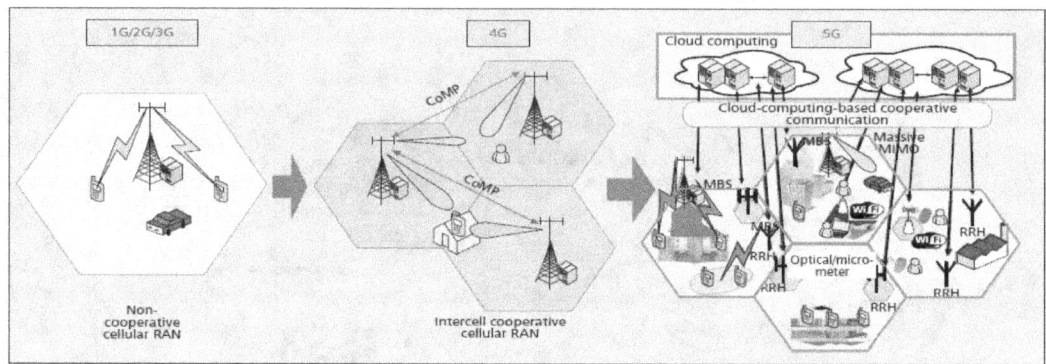

Fig. 3.22. The evolution milestone from conventional 1G to the H-CRAN-based 5G system

To optimize BBU utilization in cloud radio access network (C-RAN) between heavily and lightly loaded bases stations, the BBUs are centralized into one unit w is called a BBU/DU Pool/Hotel. A BBU Pool is shared between cell sites and virtualized A BBU Pool is a virtualized cluster that can consist of general-purpose processors to perform baseband (PHY/MAC) processing, as shown in figure 3.23. The X2 interface in a new form, often referred to as X2 organizes the inter-cluster communication. The concept of C-RAN was first offered by IBM under the concept of "wireless network cloud" (WNC) and implemented on the principle of distributed wireless communication systems [41].

The letter C in "C-RAN" referred to Cloud, Centralized processing, Cooperative radio, Collaborative, or Clean. Figure 3.24 shows an example of a C-RAN LTE network. The fronthaul part of the network spans from the RRHs sites to the BBU pool. The backhaul links the BBU Pool with the mobile core network. At a remote site, RRHs are co-located with the antennas [41].

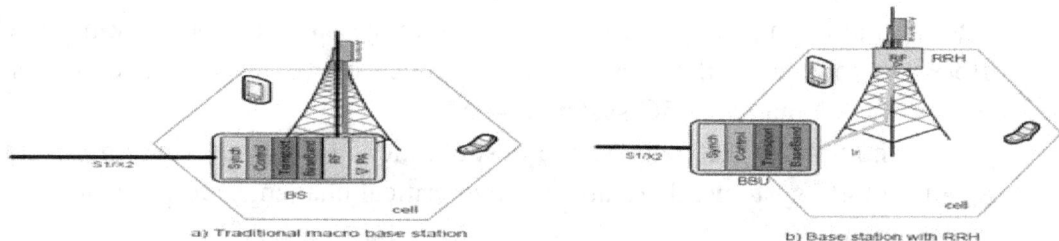

Fig. 3.23. The Baseband station architecture evolution

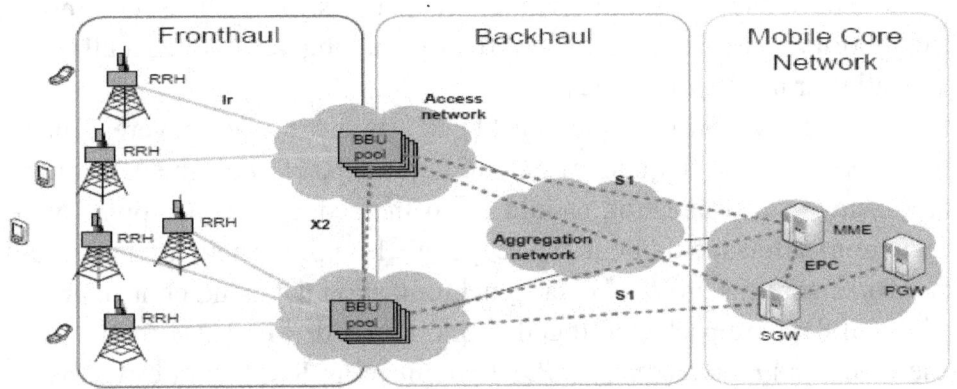

Fig. 3.24. The C-RAN LTE Mobile Architecture

3.5. Timing synchronization of LTE Network

Communication networks are continuously shifted from time-division multiplexing (TDM) based on the packet-based formation. According to this conversion, there are vast amounts of synchronous demands that require precise timing clock to be carried over the packet networks. Models of accuracy timing delicate forms that require the carrier of synchronization over packet networks involve interconnection and carrier of TDM services over packet networks (TDM switches, TDM PBXs, voice, video-conferencing, and broadband video), and joints to 2G, 3G, and 4G wireless base stations. TDM networks, unlike packet networks (e.g., Ethernet, IP, MPLS), have timing variation inherently constructed inside them. Native Ethernet (IEEE 802.3) is genetically asynchronous and was not produced with timing directions in memory. The source clock is typically obtained from TDM (E1/T1) interfaces associating the base station to the switching centers (in frequency division duplexing "FDD" networks) or from costly GPS receivers positioned at the base station (in time division duplexing "TDD" networks). GPS receivers for telecom network synchronization have a much higher designation (high-quality oscillators, high holdover capability accuracies,.) than those in the average compact satellite navigating system, plus they require all the best interfaces and cabling to interact with the telecoms equipment. The primary reference clock (PRC) is the master clock from which all other clocks in the synchronous network directly or indirectly determine their timing. This authority of time synchronization is necessary for the precise functioning of the network as a whole. A clock that ultimately drives its rate from the PRC is assumed to be identifiable to that PRC. Accurate synchronization (of not more than 50 ppb) is critical for mobile networks because the radios utilized in these networks operate in very strict frequency bands that require separation to bypass channel intervention, which decreases the call quality and network capacity. Reliable radio carrier frequencies at the transmitter antenna outputs confirm that radio channels of the adjacent cells do not overlay in the spectrum used. Spectral channel overlay makes channel cross-interference, which in turn causes noticeable and times degradation of voice quality (i.e., signal-to-noise ratio). Without timing information traceable to a highly accurate PRC, local interference between channel frequencies, as well as common interference with neighboring base stations occur, finally making abandoned calls and discrediting the overall user participation [42].

Time-division multiplexing (TDM) methods want severe synchronization that involves two perspectives: frequency and time (or phase). For the LTE network, the efficiency of frequency synchronization should be in the scope of ±0.05 ppm, while the precision of time synchronization should be in the scope of ±1.5 ppm. There are three kinds of data in any mobile network: wireless protocol data, synchronization data, and control and management data. Those kinds of data are packaged together and forwarded in TDM mode. Upon receiving the frames, the clock and data recovery (CDR) circuit of an RRH can derive the frequency information to perform frequency synchronization. Meantime, the carrier time is nearly fixed and can be estimated by BBU. Based on the determination, the timing among the BBU and RRH can be configured in approach. With the timing information derived from the frames, time synchronization at the RRH can be performed. The Ethernet-based data is no longer constant due to the packet-switched nature of Ethernet; as a result, frequency, and time synchronization among the BBU and the RRH possibly becomes difficult [43].

LTE uses two averages of radio duplexing systems: frequency division duplexing (FDD) and time division duplexing (TDD). FDD has two separate frequencies for sending and receive at the radio access point. TDD utilizes the corresponding frequency for both send and receives. While this provides for spectrum performances, there is a combined difficulty of higher phase synchronization conditions at the eNodeB, so the base stations do not conflict with each other [40]. FDD synchronization covers demands for frequency precision but may additionally combine phase efficiency with helping unique characteristics, for example, enhanced circuit switch fallback (ECSFB) or enhanced multimedia broadcast service (MBMS). For these cell sites that are within 3km of each other, TDD synchronization specifications have higher phase accuracy of 3μs to support E-NodeB alignment of 1.5μs, which is required for enhanced Inter-cell Interference Coordination (EICIC) or Up-Link/Down- Link-UL/DL coordinated scheduling for higher synchronization requirements of frequency, phase and time. The holdover period is the measure of time that an oscillator can manage synchronization (frequency and phase) inside a specified performance purpose next to the decline of its source. The essence of any synchronization scheme is its oscillator. Almost globally, LTE base station devices use oven-controlled crystal oscillators (OCXO). Depending on the OCXO performance properties, kind of crystal form, and temperature variation ends, the macro LTE base stations may afford from 4 to 18 hours of holdover with an accuracy of ±1.5μs [44]. The difficulty of synchronization in the LTE network is the clocks from various network devices and many protocols that were invented in the past. One of the most widely used ones is the network time protocol (NTP). It is applied all over the Internet and based on a hierarchical construction of time servers where each server receives the reference time from its parent and publishes it to its children. Due to this structure, it scales well, but the precision of the synchronization suffers on every hierarchical layer of the structure [45].

One answer is to practice an adjustable frequency oscillator to calibrate time in the RRH, which can correct the oscillator frequency periodically to guarantee a continuous-time difference. When it occurs to maintaining MIMO or TX variety transmission technologies, the time synchronization specification is severer and should be in the scope of ±65 ns. More attempts are required to understand how to meet the ±65 ns time synchronization specification, such as expanding a GPS antenna on the BBU, developing the precision of timestamps, rising clock frequency, and optimizing the variation improving algorithms [39]. There are different proposed solutions for transferring the clock signal inside the IP network of the LTE; one of these solutions is using the precision time protocol (PTP), version 2, and IEEE 1588 which can offer precise synchronization of clocks over heterogeneous systems with accuracy in the microsecond to sub-microsecond range. The supported protocols for the PTP system are UDP, IEEE 802.3 (Ethernet), Device Net, ControlNet, and PROFINET. Each PTP entity has one or more ports, communicating with other clocks in the network. The PTP standard defines several clock types [44]:

The master clock: It considered as the source of time; many clocks on a particular path synchronize with it.

- The ordinary clock: It has a single port, and it works as a slave clock it is also synchronized with a master clock
- The boundary clock (BC): It has multiple ports. It acts as a master and slave clock. Usually, it does not associate with the application domain devices (e.g., actuators or sensors)
- The transparent clock (TC): it considered as a critical clock type to meet the required specifications, such as providing microsecond synchronization requirements over multiple network nodes from a given slave clock to the master clock. Specifically, a TC is a device capable of measuring the residence time in a router/switch, which includes, for instance, the measurement of processing and queuing delays. The residence times are taken into account to measure the end-to-end delay accurately at the slave clock.
- The peer-to-peer transparent clock: It used to measure not only the residence time information, but also measures the link propagation delay between similarly equipped ports at the other end of the link.

As listed above, the transparent and peer -to- peer clocks require specialized PTP hardware at intermediate nodes. As opposed to the network time protocol (NTP) does not need any particular device between the slave and master nodes. At each slave node, the master clock protocol is executed such that every clock synchronizes with the best clock as each slave node examines its local ports to find the announcements of the clock messages and selects the best one as a master and put the others in lowest priority in terms of clock class, and accuracy. Note that the slave node will recalculate and reselect a new master clock according to the highest priority in case of a failure of the master. During the time of the reselection process, the slave nodes are running without any synchronization clock. It is proposed to maintain the same quality of the latest master clock and synchronize using a freezing algorithm for improved convergence time [43].

The main steps of the PTP request-response protocol, as follows:

- The master clock sends a Sync message periodically to a given slave node, this message contains the send time T1, and the rate of Sync messages is typically set to 1-2 seconds.
- Depending on the timestamp algorithm (with or without hardware support), a follow-up message containing T1 is sent.
- Once the message containing T1 is received by the slave node, the receive time T2 is recorded.
- Then the slave node sends a delay message at T3 to the master node.
- The master node notes the reception time T4 and sends it back to the slave node in a delay message.
- After receiving the delayed message, the slave node is aware of all required timestamps (T1; T2; T3; T4), which are used to correct its local clock.

One significant difference between PTP and NTP is that the overall process is initiated by the master clock in PTP. Also, PTP contains algorithms to enhance the time synchronization accuracy, which is covered in the following [43]:

- Delay and offset approximations: the PTP offer three techniques to reduce the adverse effects of asymmetric links on the synchronization performance as following:
    - Residence time at intermediate nodes (TCs and BCs): The residence time at a given router/switch consists of the time duration a PTP packet resides in the switching fabric from the input port to the output port.
    - Asymmetric delay parameter: If the fixed symmetric delay properties are known for a given connection, an asymmetric delay parameter can be used, corresponding to the delay asymmetry field in IEEE 1588. It is worth noting that the delay asymmetry field is added in the correction field, where the latter field integrates all correction delays.
    - Peer-to-peer path correction: Peer-to-peer transparent ports measure the link propagation delays and include this in the correction field. This technique helps to better estimate the delay at the cost of an increased overhead frame.
- A slave clock synchronizes itself with the master clock following the PTP techniques, and at the time, one intermediate node has a transparent clock and measures the residence time. For the nodes which have no PTP support, such as the regular router. It is recommended that it processes the PTP messages with high priority to decrease the residence time. For improved accuracy, all nodes in the example should have transparent clocks, as shown in figure 3.25 [43, 46].

Another study of the new synchronization standard is the synchronous Ethernet (SyncE), standardized by the International Telecommunication Union (ITU) in ITU-T G.8262. Note that the native Ethernet protocol works only between two nodes without any need for synchronization, where the clock accuracy is 100 ppm. Effectively, the preamble in Ethernet frames allows a physical layer in the receiver to start a new data transmission and extract the bitstream directly. The fundamental synchronization principle, of the synchronous Ethernet, consists of transmission the clock signal at the physical layer of the Ethernet and propagating it to all Ethernet nodes with the support of the synchronous Ethernet. The slave acts as a timing master once a reference signal is received. The recovered clock at each intermediate node is recovered with a phase-locked loop (PLL), and then it is passed to a clock multiplier unit (CMU), which generates the output clock. The PLL attenuates the accumulated clock jitter. Timing devices in a synchronous Ethernet network should support 4.6 ppm in free-run, according to the worst-case performance scenario. Further, synchronous Ethernet acts on a hierarchical master-slave manner [43].

The synchronous Ethernet node generates the reference clock and transmits the clock signal to its adjacent hob slave port. The slave port then sends the signal to the transceiver, which works as a master port. As the reference clock signal becomes propagated over all adjacent synchronous Ethernet nodes, the overall nodes of the network get synchronized. Also, the physical layer sends dummy data continuously in parallel with the regular data transmissions, as the same technique of the Synchronous Optical Networks (SONET) [44]. It is terrible to notice that all hobs in the network path must support synchronous Ethernet to be synchronized. Therefore, synchronous Ethernet does not offer Time-of-Delay (ToD) synchronization, which is a crucial feature of the PTP protocol. The synchronous Ethernet and PTP are complementary; where synchronous Ethernet is used to provide tight synchronization, and also to improve the PTP performance, where PTP can use frequency timing inputs from synchronous Ethernet. The combination of synchronous Ethernet and PTP was used in a telescope for nanosecond timing resolution. Furthermore, the white rabbit synchronization system, described in the following, also uses PTP and synchronous Ethernet in particle accelerators and detectors, as shown in figure 3.26 [42]. Another way in the synchronization of the IP network called white rabbit (WR) synchronization is to combine the PTP and synchronous Ethernet standards to meet sub-nanosecond accuracy requirements for particle accelerators and cosmic particle detectors; these applications require deterministic and reliable data delivery as well as sub-nanosecond synchronization accuracy.

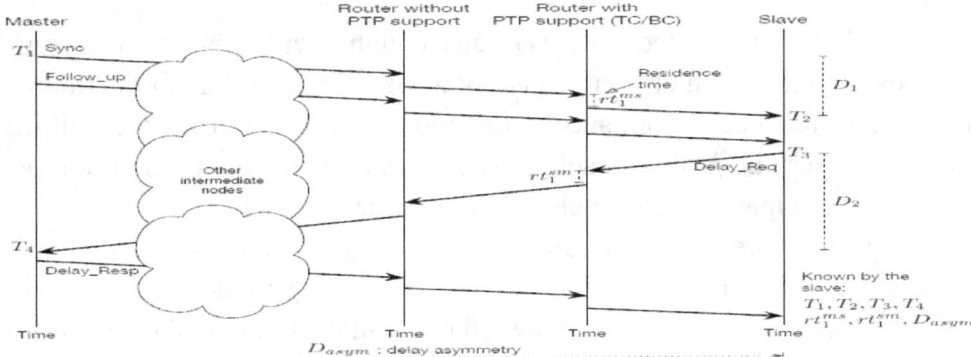

Fig. 3.25. The PTP synchronization protocol

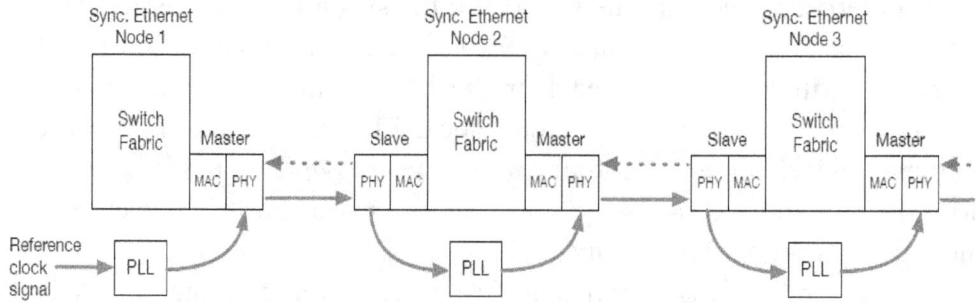

Fig. 3.26. The Synchronous Ethernet

A key feature provided by PTP used in the white rabbit is the residence time, which is a measurement in the network and takes into account the average node variable delays. The white rabbit detects asymmetry synchronization, which is no detected by PTP to enhance synchronization accuracy. Some of the analytical techniques proposed by the white rabbit are as following [42] :

- The White Rabbit link setup: Before executing the regular PTP protocol, an initialization protocol works for node identification, synchronization, and for providing the parameters of the link.
- The White Rabbit link delay model: The delay from a master to slaves is calculated according to fixed and variable parameters: Dtx+dms+Drx; where Dtx and Drx are fixed delays for the transmission circuitry at the transmission and reception sides, respectively, and DMS is a variable delay of the transmission medium considering the fiber refractive indexes.
- Synchronization using synchronous Ethernet: For enhanced synchronization, synchronous Ethernet is used between master and slave nodes to share the same frequency.

3.3 Conclusion

Applying the 5G technology in any country will depend mostly on the solutions of the previous challenges which were faced by the mobile operators in the traditional 4G network. To achieve an excellent performance in the communication market the builders of the 4G network should consider the proposed solutions of these difficulties in the traditional mobile and apply the cloud principles on all levels of the mobile network, at same time all the network parts including the core, metropolitan and the access network should work under the same management system to avail the required flexibility to develop the existing network at the different levels of the infrastructures. The virtualization of the network is the secret key to achieve the required improvements in the communication network, and these will make the concept of centralizations in all tasks of the network such datacenters more comfortable to use and the availability to apply the concepts of the cloud network at the levels of radio access and IP core network will be exist.

Chapter 4

Intelligent Security of the 4G Traffic over the Optical Transport Network

4.1 Introduction

The transportation of the client data signals across the global communication networks is considered a critical issue to most of the network operators in how they can keep the confidentiality of the contents of these data through its complete journey in a global network, also in how they can avail the network resources to carry and maintain the performance of a considerable amount of data bits from the source to the destination. The Optical Transport Network (OTN) technology was designed to transfer multiple types of clients' signals with different data rates through different wave-lengths and over the same fiber optic cable by using Dense Wave Division Multiplexing (DWDM) technology. The process of transporting the client data signals from the source to the destination in the core transmission network includes many steps to map these signals into the generic OTN frames which depend on the data types and client data rates, where the client data is received at the client port in the Network element of the optical transmission network, the transponder card encapsulates these signals in the payload container and at the same time it adds the required overheads to perform the optical payload unit (OPU), the next step it multiplexes many of low payload rates to higher optical data units (ODU) rates, and the final step in the mapping process is to perform the optical transport unit (OTU) frames of the optical network. The security of the client data while it is traveling journey across the OTN of the core Transmission network depends only on the standard encryption algorithms of the client signals at the application layers. These signals travel through different layers in the optical transmission networks, and the operators of these networks rely mostly on the security algorithms of the client signals at the application's layers only.

However; the problem with the majority of the optical transmission networks is that an attacker can access the physical layer from the optical fiber cables and split the optical signal by wiretapping it, after that he will be able to keep a live copy of the optical signal, and by trying different types of the reverse engineering in the de-mapping processes on this live copy according to the standard structure of the OTN frames, the attacker will reach to the original layer of the client signals. On the other hand, by using different types of the decryption algorithms in the service layer, the probability that the attacker can reach to original contents of the client data will be very high, and he will be able to break the encryption technique in the optical network from the first try by 100% percent success [3].

The necessity to secure the essential client data signals while it was traveling through the optical transmission network remains crucial, especially on the optical cables which are passing through long routes, and are vulnerable to be split and wiretapped by the interested attackers from any undetermined place [47].

Another significant attack that targets the optical core networks is the jamming attacks; this attack is made by accessing the optical fiber links from any place in the network, and additional harmful data signals are inserted inside the contents OTN frames. The jamming attacks are to make service degradations, and the results are misleading or modifying the original data contents of the client data signals. Most of the previous studies conducted many solutions to the security problem of the physical layer in the optical network ,one of these studies proposed to make security profiles for every user in the optical network and creates upper category from these users which called gold users , where the gold user will have 1+2 protection links against any security attacks in the physical layer of the optical network, as an example for every user link there are 2 other extra links, and the user has the ability to reroutes his traffic on these extra links in case of the original link is not safe , this solution is very costly and wastes the resources of the optical network especially in case of many gold users [48]. Other studies discussed that how to use the XOR and optical LFSR to secure the optical data signals despite the difficulties on how they may face the distribution of the encryption keys between the different NE's in the optical network [49, 50]. More studies discuss new algorithms for the optical data encryption that uses quantum noise inherent in laser light and modulations algorithms that use two different cycles of the M-ary phase-shift-keyed (PSK) signal, a technique which allows the receiver who owns the short secret key to transform M-ary signals to standard modulated signals. The attacker, who does not own the secret-key, will be forced to try different M-ary measurements several times, which is very difficult [47, 51]. Despite there were many numbers of studies proposed many years ago on optical security, the physical layer of optical networks is still the weakest loop in communication networks from a security perspective overview.

In this chapter, a new proposed model for securing the client's data over the optical network, and this will be done by adding new security layer through extra separate module in the NE's, which will be used to make the required functions of the encryption algorithms for the selected clients' signals as per customer choices. By adding this security layer for the different client frames in the OTN structure, the wavelengths over optical cables will be more secure from any unauthorized access and wiretaps hacking. On the other hand, implementing security algorithms in a separate stage in the OTN system for certain client signals only as an option will reduce the complexity of transmission systems, especially in the optical networks with huge capacity. , for the first time, the new proposed model is studied about how to secure the customer's data over the optical system with the help of the artificial intelligence (AI) prediction feature. The training data set of the artificial neural network (ANN), which used to predict any intrusion in the physical layers are organized according to the dynamic variations in the optical signal to noise ratio over the entire network. The response to any intrusion detections in the optical network will trigger from to enable a new security layer through the mapping stages of the OTN frame in the NE's, and it will be utilized to perform the required encryption algorithm for the customers' signals according to administrator decision. By including this security layer for the distinctive customer in the OTN structure, the wavelengths over the optical links will be more secure from any unauthorized access and wiretaps hacking. All the encryption algorithms in the proposed security layer will be built by using XOR gates and LFSR, and by considering that the client signal 10/ 100Gb/s is equal to the same capacity of the wavelengths in the DWDM systems. The proposed security layer will be enabled according to the results of the intrusion detection phase in the optical network. The detection of any intrusions in the network will be done by using centralized security controller (CSC) for all the optical network elements, the main job of the CSC is to monitor the variations of the optical to signal ratio (OSNR) of the target links in the optical network. In case of any up-normal changes in the OSNR values in one link of the optical network, it will be processed according to the risk of new attack in this link will be stated, and according to the investigations of the network operator about these changes the CSC will enable the security layer for the client data signals which travel through the intruded link or it will make reroute to the services to other links until the network operator finishes the investigations about root cause of these changes (See Appendix A).

4.2 The Structure of the OTN Frames in the Optical Network

There are different mapping stages of the client signals inside the NE's of the transmission network consist of:

Stage1: The transponder card which receives the client signals with rates 10/100 Gb/s transforms it to the electrical domain by using Small Form-Factor Pluggable (SFP) module with a suitable laser frequency for the client signal. The next step in this stage is the mapping of the client signals to be part of the structure of the OTN frames, and this is done by putting the client data in containers with fixed size according to the client data capacity. These containers are used to form the optical payload units (OPU_k) (where k represents the capacity of the OTN frame with k = 2 represents a frame with data rates equal to 10 Gb/s and k = 4, OTN frame with data rates equal to 100 Gb/s) by adding its overheads, the next step is to form the optical data unit ODU_k. Finally, on adding the final overheads and alignments words with forwarding errors corrections (FEC), the optical transport unit (OTU_k) will be formed as shown in figure 4.1, and by converting the OTU_k to the optical domain it will ready it to be multiplexed as a wavelength in the WDM transmission system [52]. The final frame of the OTU_k in the OTN system is shaped in 4 different rows of 4080 bytes and repeated every certain period (12.191 usec for OTU_(2) frame and 1.167 usec for OTU_4 frame) [53]. This frame consists of bytes that stand for frame alignment word (FA), different layered overheads bytes (OH), payload data, and finally bytes that represent forward error correction algorithms. After that, this frame is converted to standard optical signals with standard wavelengths, as shown in figure 4.2 [54].

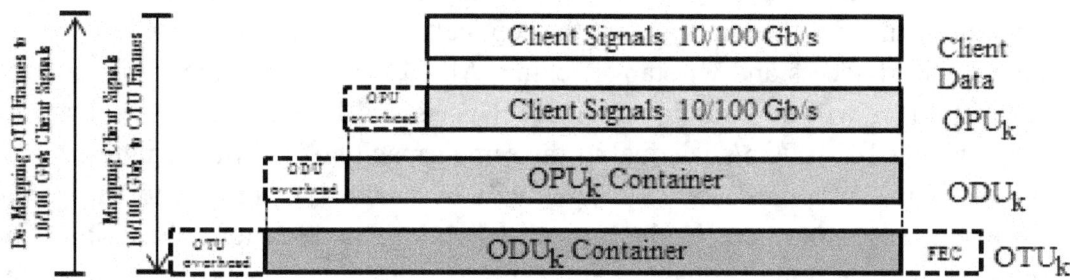

Fig. 4.1. The Construction of the OTN

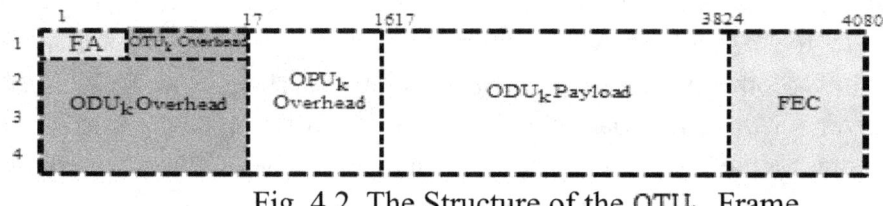

Fig. 4.2. The Structure of the OTU_k Frame

Stage2: Giving that the network consists of two network elements with two directions in the optical network, the 2'nd step in the optical transmission system processes is the multiplexing of many standard optical signals to form one beam of laser which consists of many wavelengths and different standard frequencies in the c band.

Stage3: The last stage in the transmission network inside the NE's is the amplification process that uses any optical amplifiers such Erbium-Doped Fiber Amplifiers (EDFAs) with the suitable gains according to different factors on the network such as the traveling distance between the two network elements and the attenuations on the available fiber cable.

The transmission model in this chapter consists of two network elements connected by a fiber cable with 40 wavelengths (W1 to W40) in the DWDM system and client signals with different bit rates, as shown in figure 4.3.

As shown in figure 4.3, the problem in the current model from the security perspective can be explained as follows:

Assume that Alice, in location A wants to send a message to Bob, in another location B Where A&B are connected through the pervious optical transmission network.

Then, the attacker Eve in location C can split the laser signals of the fiber cable from any point of the transmission network into two paths by using optical splitters where the first path of the laser signals will be returned to the original route of the network to transfer original message of Alice to Bob and the 2'nd path will be used by Eve's equipment where Eve can have a live copy of the OTU frames without affecting the original routes of the network.

and in case Eve's hacking system made a little effort in the de-mapping process according to the standards of the OTN structures, Eve will be able to break the optical network structure and will know the contents of the messages between Alice and Bob while they do not realize that, all their communications messages are known by the third person who is Eve.

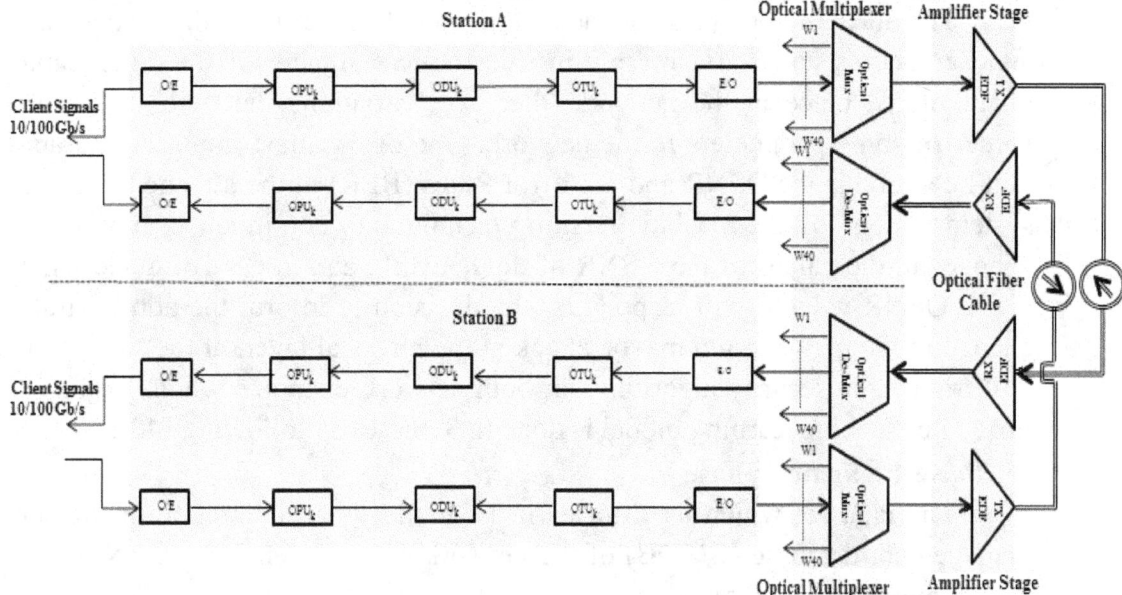

Fig. 4.3. The Transmission Model of Two Network Elements

As shown in the previous optical transmission network, there are not any security algorithms that are implemented to prevent the interested attacker such Eve from understanding the contents of the optical signals even though he can access the fiber cable and keep a live copy of the client signal

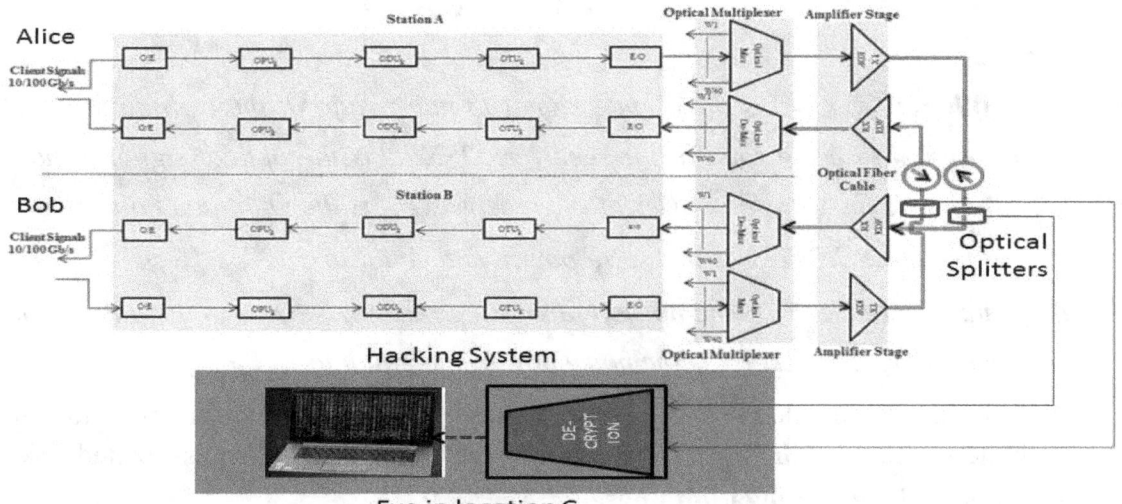

Fig. 4.4. The security problem of the Transmission Model

4.3 The Proposed system model

4.3.1 Intrusion Detection and Response Model with machine learning

The idea of the intrusion detection model is built on that when any interested attacker access and starts in splitting the optical signals of the fiber cable from any unauthorized place in the network, there are variations that will be happened on the values of some parameters in the network. One of the most important parameters which will be varied is the OSNR and Bit Error Rates (BER) in the attached link. The proposed system is done by the machine learning technique to continuously monitor the dynamic changes in the values of the OSNR of the network, and in case of detecting any changes in the OSNR value it will respond by the decision to inform the administrator there is a change in the network and maybe attack at the physical layer, at the same time the model will trigger the security algorithms in both network elements which contain the infected link. The machine learning model is done in 3 phases as following [48]:

- **Phase 1 learning phase**

The data set which used to learn the model is an extract from the server change management database (CMDB) of the network management system (NMS) according to the following parameters:

$$\theta_{att} = r \tag{4.1}$$
$$\theta_{rx} = 1 - r \tag{4.2}$$
$$OSNR_{Tx} = \frac{P_T}{P_Q F_T} \tag{4.3}$$
$$OSNR_{Rx} = \frac{P_R}{P_Q}(\frac{P_N}{P_Q} + \theta_{rx} F_R - 1) \tag{4.4}$$
$$\frac{OSNR_{Rx}}{OSNR_{att}} = \frac{\theta_{att}}{\theta_{rx}} = \frac{r}{1-r} \tag{4.5}$$

Where: $OSNR_{Tx}$ is the optical signal to noise ratio at the transmitter, $OSNR_{Rx}$ is the optical signal to noise ratio at the receiver, $OSNR_{att}$ is the optical signal to noise ratio at the receiver, P_T is the transmit power at the receiver, P_Q is the Quantum noise power and equal to -58 dBm at 12.5GHZ, P_N is the noise power and equal to $P_Q F_{T(R)}/G$, F_T is the noise figure at the transmitter side, F_R is the noise figure at the receiver side, G is the gain of the amplifiers at receiver or Transmitter, P_N is the noise power and equal to $P_Q F_{T(R)}/G$.

As the small value of r, the model will not be able to detect the intrusions in the network; with the increased value in r the system will detect the intrusion in associated link

- **Phase 2 the generalization phase**

In this phase the evolution of the model will be done according to many examples of data set which is not used in the learning phase and the errors will be calculated by using the root mean square errors (RMSE) will be used as the following:

$$RMSE=\sqrt{\frac{1}{N}\sum_{k=1}^{k=N}(\hat{r}_k-r_k)^2} \qquad (4.6)$$

- **Phase 3 the implementation phase**

The implementation phase for the simplicity is done by using three-layer artificial neural networks (ANN) With the following variables [55]:

Layer1 is the input variables such as transmit power, receive power, OSNR values at transmitter and receiver, the distance of the span, the total attenuations

Layer 2 is the hidden layers with the neural cells which will be trained as in phase 1 to find the relations functions between the output and the input

Layer 3 is the output layer which expresses the prediction of the intrusion on the optical fiber cable

Layer 4 is the response layer with many output responses as it enables the security layer algorithm in the infected section. Figure 4.5 shows the ANN proposed model with three layers only.

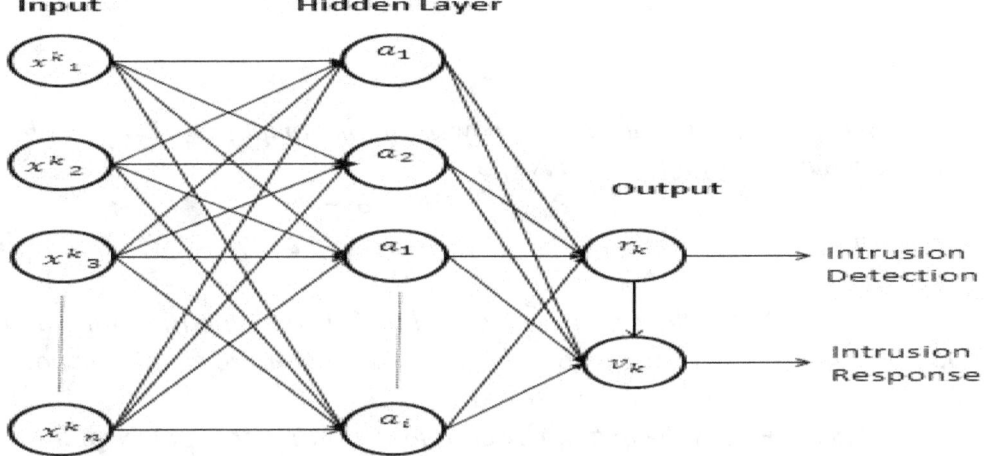

Fig. 4.5. The ANN Model of Intrusion Detection and Response

As indicated before, according to the decision of the previous ANN model, anew security layer will be empowered to the customer data over the OTN frames in the DWDM transmission system from any wiretapping attack. It is recommended that the security layer be actualized in the ODU_k Bytes, as shown in figure 4.7.

4.3.2 Intrusion Detection in the physical layer of the optical network with the software-defined network:

In the recent years the software-defined network (SDN) technologies attracted many researchers to apply its concepts on many fields in the network, the most advantages of the SDN concept is it transforms the control parameters, the protocols and the functions of any network to be programmable and this done by separating the control plane from the data plane. Extending the principles of the SDN to the optical transport network can provide a new framework in the applications, coordination, and orchestrations of the higher-order optical layer [56].

In our security model the principles of the SDN in the intrusions detection and requested response in the physical layers of the optical network, and this done by continues monitoring the variations in the optical signal to noise ratio (OSNR) in the different optical links by using proposed software-defined security (SDS) for all the physical layers of the optical transport network. In the SDS model, the control plane is separated from the data plane by using a centralized Security Controller (CSC) to perform this function in all the optical sections. The CSC measures the values of the OSNR automatically for the different optical links by monitoring the performance of these links from the change management database (CMDB) of the operation support system (OSS) server and the different network management systems (NMS's) of the optical transport network. The OSNR values for every link can be calculated according to the following equations [55]:

$$OSNR\ (DB) = P_{out} - P_{ASE} \tag{4.7}$$

Where: P_{out} is the output power of the amplifier, P_{ASE} is the total amplified spontaneous emission (ASE) noise power.

$$P_{ASE} = NF + G + N_{in} \tag{4.8}$$
$$= 10\ log\ 2\ n_{sp} + 10\ log(G-1) + 10\ log\ hvB_o$$

Where: NF is the external noise index, G is the amplifier gain, and N_{in} is the input noise power $N_{in} = 10\ log\ hvB_o$ which means input noise equals the power of photon.

$$G = P_{out} - P_{in} \tag{4.9}$$

Where: P_{out} is the output power, and P_{in} is the input power of the amplifier

Figure 4.6 shows the proposed model for the SDS model which performs the function of monitoring the variations of the OSNR in the optical network, and this done by using centralized security controller to perform this function only as of the following sequence:

- The operation support system (OSS) of the transmission network collects the performances of the optical links from the Network management layers (NMS's)
- The centralized security controller (CSC) communicates with the change management database (CMDB) of the OSS server and checks any variations of the performances of the mentioned optical links
- In case of any changes in the performance, the CSC will start and calculate the OSNR of this link after that it will compare the results with the threshold values of this link
- If the values of the OSNR are less than the threshold values for this link, the CSC will send notifications to the network operator about these changes
- The network operator will decide if these changes are determined by defined root cause or not determined

- If the root cause is unknown of these changes, the CSC will enable the security layer for the selected traffic which passes through the defected link

4.3.3 The Proposed Security Layer

The customer signals are mapped into the OTN frames as indicated by its bit rates, after that at the ODU_k level the data is scrambled by the proposed security model and the encrypted data (Encrypted ODU_k) is returned to fill the frame of the OTU_k In the OTN system. The proposed model consists of the intrusion detection in the physical layer of the optical links and the security model which performs the cryptographic algorithms to the client data signals after the detection of the intrusion in the optical link and according to the previously proposed model the CSC will enable the proposed security model to secure the client signals over the OTN system at the source and destination NE's only in the DWDM transmission network. The security techniques will be against the wiretapping of the optical signals and are done by adding a new security layer in the mapping process of the client signal and before forming the final frame of the OTN system. According to the proposed model, the security layer is implemented after forming the ODU_k bytes of the frame and before the final OTU_k stage, as shown in figure 4.7.

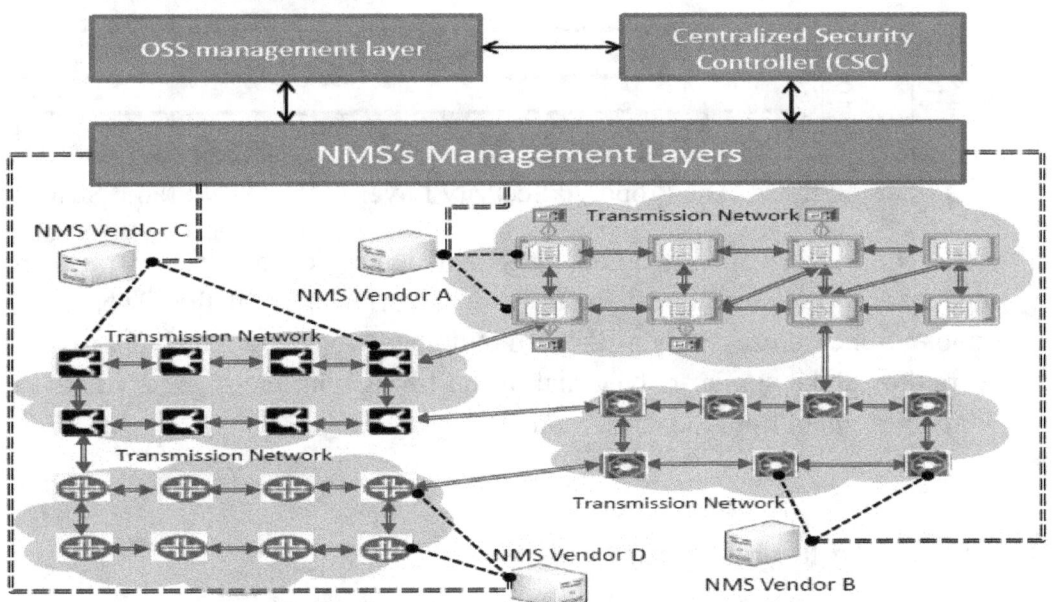

Fig. 4.6. The proposed Model for the intrusion detection

To achieve this security layer, there are two types of Pseudo-Random Generators (RNG) that could be used to generate the keys of the encryption/decryption processes. The 1'st one is synchronized RNG's while the 2'nd is an unsynchronized RNG's, provided that both types of the RNG's systems are working with the same linear feedback shift register (LFSR) and by using XOR operations to perform the encryption/decryption processes , the only difference between them is that in the first type of the RNG the source and destination stations are synchronized with the same clock and it generates the same keys in both stations as result of using the same clock for synchronization, While in the 2'nd type of the RNG the source and destination stations are working separately in every station with different keys generation. There are two sorts of the Pseudo Random Generators (RNG) that could be utilized to create the keys of the encryption/unscrambling forms. The first one is synchronized RNG's while the 2'nd is an unsynchronized RNG's, the two kinds of the RNG's frameworks are working with the Linear Feedback Shift Register (LFSR) and by utilizing XOR activities to play out the encryption/unscrambling forms , the main contrast between them is that in the primary sort of the RNG the source and goal stations are synchronized with a similar clock, and it produces the equivalent keys in the two stations as the effect of utilizing a similar clock for synchronization, While in the 2'nd sort of the RNG the source and goal stations are working independently in each station with various keys age [53].

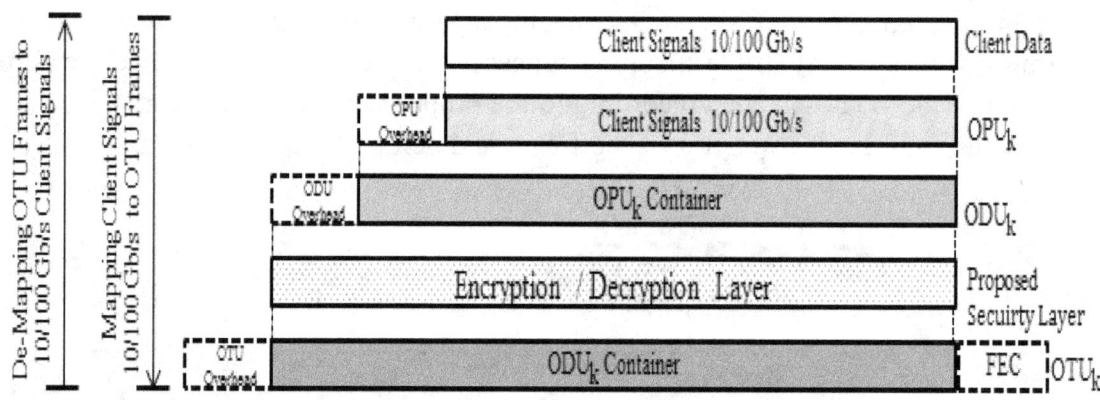

Fig. 4.7. The Proposed Security Layer in the OTN Frame Structure

The secret key generation rates will depend on the trigger rates from the ANN model. The proposed model for the new security layer of the OTN frame is to utilize double Synchronize Pseudo-Random Number Generator (SRNG) to create a similar secret key and a similar polynomial of the LFSR with the trigger with every clock pulse which generated from the ANN model.

Generating the Encryption Key

In our model, the security over the entire optical network is managed by the centralized security controller (CSC), which is part of the software-defined security (SDS) of the optical network, as shown in figure 4.8. As soon as the CSC discovers any intrusions in optical links, it will transmit the trigger pulses and the initial symmetric key to the source and destinations NE's of the selected client data signals. The transmission of the initial keys to the NE's is performed by using secure transmission lines, and the generation of the initial symmetric keys by the CSC may include Cryptographic Secure Pseudo-Random Number Generator (CSPRNG). The source and the destination NE's will initiate its security model by using the initial symmetric key which was generated and transmitted before by the CSC, and the proposed security model starts to generate its encryptions- decryptions keys according to the combinations between the data plain bits and the initial key [57].

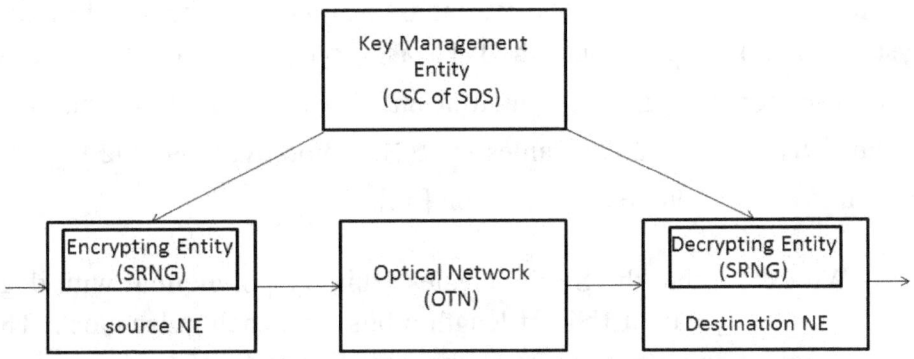

Fig. 4.8. The Key Management Entity in the optical network

For generating the primary key, a chaotic map has been used with the following equation:

$$f(x_i) = \begin{cases} x_i/\gamma & x_i \in [0, \gamma] \\ (1 - x_i)/(1 - \gamma) & x_i \in (\gamma, 1] \end{cases} \quad (4.10)$$

Where: γ is the control value x_0 is the initial state, and their value included in the interval (0, 1).

A critical characteristic of the selected chaotic map is it has no periodic widows and maintains the chaoticity in the whole parameter space, as any selected key will meet the required chaoticity according to space the cryptographic system parameters. One weak point of the chaotic map is its digital applications depends on finite word length, to overcome this disadvantage point of the chaotic map in our model we use the Pseudo-Random Number Generator (PRNG) as it disturbs the chaotic orbits and the randomness of this system can be improved [58].

The Proposed Security Model Implementation

The proposed security layer is performed according to the traditional techniques of the Encryption - Decryption processes as the following concepts [59]:
- Consider the length plaintext of the data as m, the output sequence of the security layer as y, and by using the secret key for the encryption operations as k

- processes of securing the plaintext m at the source station are done according to $y_i = m_i \oplus k_i$ where i is the bit number in the plaintext frame and $\oplus$ is the XOR operation To retrieve the original plaintext at the destination station, the decryption process will be implemented by using the same secret key according to $m_i = y_i \oplus k_i$

The transmission over the optical networks takes place by transmitting the bits within periodic frames with different frame rates (in the synchronize digital hierarchy system (SDH) is 125us, and the OTN is varying from 12.191 usec for OTU_2 frame to 1.167 usec for OTU_4 frame) which means that for every 1 second there are 8000 frames in the SDH system, 82027 frames for OTU_2 moreover, 598802 frames for OTU_4 will be transmitted from source to destination [52].

We expect that the SRNG creates arbitrary polynomial with degree n and random generation key for the LFSR of length n bits with each pulse clock. The proposed model for the encryption layer appears, as shown in figure 4.9 [52].

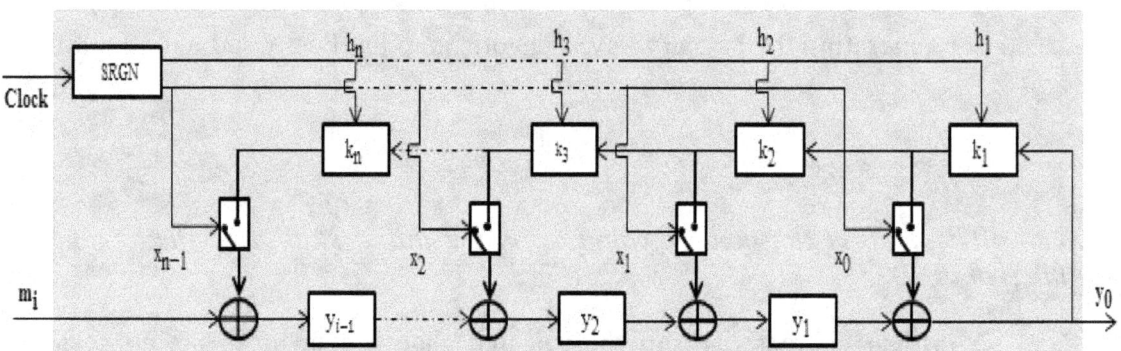

Fig. 4.9. The Proposed Model of the Security Layer

The proposed model consists of:
- The SRNG which generates random polynomial $P_n = f(x_n)$ with degree n, and used to enable or disable the switches in the proposed security layer and at the same time generates a random key k_n with a stream of bits h_1 to h_n as well as to initialize the LFSR with every reset cycle of the encryption period.
- Flip-flops to keep the encryption bits with every piece move encryptions
- XOR to implement the encryptions/decryptions algorithms.
- Plaintext which is the information of the ODU_k moreover, equivalent to $M = N * m$ where N is the quantity of the reset cycles and m is the length of the piece of the plain content

which will be encoded/decoded by the same initialized key and the generated polynomial

- Encrypted output data y_i.

As assumed, the SRNG selects one polynomial according to the following equation [60]:

$$P \leq \psi(n-1)/n \quad (4.11)$$

Where $\psi(.)$ is Euler function, the polynomial degree n and restricted by $n_{min} \leq n \leq n_{max}$, with n_{min} is equal to the key length which is used to encrypt data message m, the original message M which is equivalent to ODU_k in every frame with a length L_M will be isolated to many equal information parts with a length $L_m = L_M / N$ and N is the number of the reset cycles to encrypt/decrypt the messages with a length l_m as per the following condition [60]:

$$n_{min} \geq (\frac{L_M}{N} - 1) \quad (4.12)$$

The vulnerability of the created polynomial, and the secret key from the SRNG is estimated by the entropy of the likelihood of right polynomial might be distinguished as the following condition [61]:

$$H_1 = \sum_{i=0}^{n} P_I (log 1/P_I) \quad (4.13)$$

Where P_i is the probability to figure one polynomial of degree n out of 2^{n-1} polynomials, and the entropy of the probability to figure the instated LFSR as per the following equation:

$$H_2 = \sum_{i=0}^{n} k_i (log 1/k_i) \quad (4.14)$$

Where k_i is the probability of distinguishing one secret keyword of length l_k out of 2^n Keys. The generated polynomial can be selected randomly with every clock pulse from 2^{n-1} Polynomials, which can be stored in buffers with the required size.

The implementation of the proposed model in the OTN transmission system in our case is done between 2 NE's source and destination, as shown in figure 4.8 and 4.9 [55, 56].

where the client data signals are mapped into the OTN frames according to its bit rates, the ODU_k data is encrypted according to the proposed security model of figure 5 and the encrypted data (Encrypted ODU_k) is returned to fill the frame of the OTU_k In the OTN system, after that, it will be multiplexed optically with other wavelengths, and the amplifier stage will be done over the optical fiber cables to reach the destination station. On the receive side, the same processes will be done but in reverse actions. After the preamplifier stage in the receive direction, the de-multiplexing stage will isolate the OTU_k also, forward it to the de-mapping operations until it reaches the encrypted ODU_k, then, the decryption processes will start searching for the original ODU_k by using the same model of the encryption system, as shown in figure 6, whereas the dual SRNG polynomial with the same secret key in both sides of the source and destination stations is used.

4.4 Analysis of the Proposed Security Model

There are many security mechanisms used in the optical transport network, one of these mechanisms is securing Only the Perimeter at the client-side, in this mechanism the client data are protected by intrusion provision system (IPS), customer edge switches (CES) or firewalls at the client equipment only. While the most vulnerable points in the network are the physical layer of the optical transport network, where this mechanism leaves the internal optical links open to any security threat. The other one of the security mechanisms in the optical network is the distributed and Uncoordinated Security mechanisms, this technique uses different and independent security algorithms which are applied for the different sections in the optical network, the mechanism increases the complexity and the need resources of the security management system, and reduces the optical network performance [58].

In case of the intruder has the ability to manage and access the fiber cable by splitting the optical laser signals and keep a live copy of the OTN frames to start the de-mapping process, the intrusion detection model will detect any change in the OSNR value of this link and immediately the system will enable the security layer by sending the trigger clock to the SRNG in both sides automatically, the attacker will only get the encrypted ODU_k instead of the original ODU_k which included in the OTN frame after the response of activating the proposed security layer.

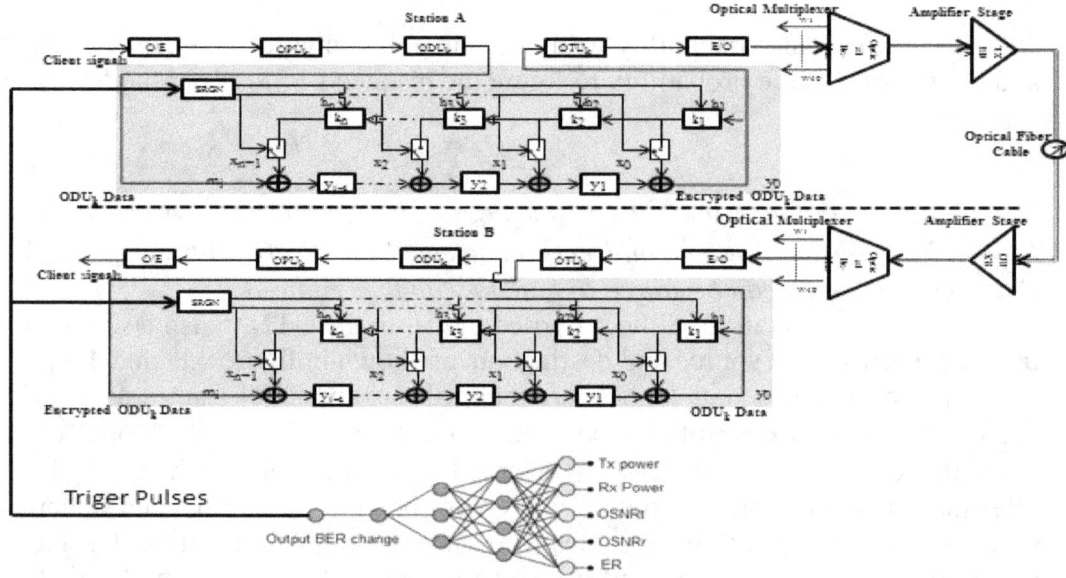

Fig. 4.9. The Implementation of the security layer in the OTN system with the machine learning model

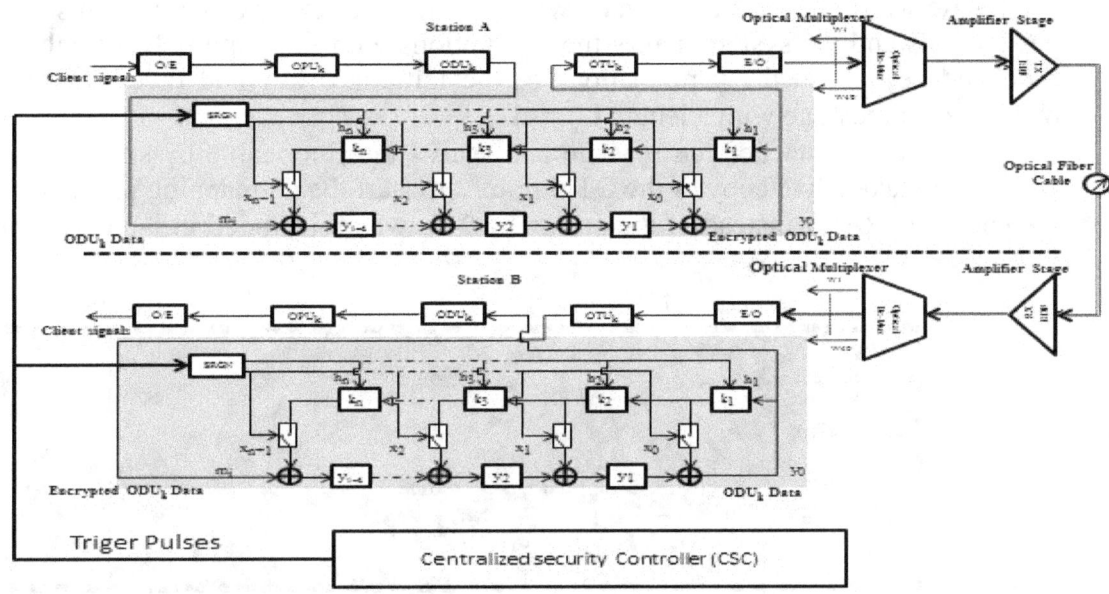

Fig. 4.10. The Implementation of the security layer in the OTN system with the SDN Intrusion Detection

In our model we used the concept of software-defined network security (SDS) for applying the security mechanism in the optical network by implementing centralized security controller (CSC) over the entire optical transport network, the CSC not only used for monitor the security breaches and makes the required security decisions but also it optimizes resources which utilized for the security algorithms in the optical network.

By using the software-defined security concept in the optical network, it centralized the security policy management, the coordination and eliminated the dependence on the vendors' security mechanisms. Having only one centralized controller in the control plane in the SDS of our model is considered weakness point from security perspective, where implementing only one CSC for all the optical network will increase the risks of the attacks on the links between this controller and the switches of the SDS of the optical network, and this may isolate the CSC from acting its function in monitoring and provides the required security decision to the different NE's in the optical network. To overcome this weakness point in our proposed model its suggested to use 2 or more CSC in the SDS of the optical network, the first CSC is set as active working controller in the SDS network and the other one is as idle protection controller, the 2 CSC's in the network should be synchronized and should exchange its data between each other and this done by using a certain types of the data replications between the 2 servers of the CSC's.

With every variation in the OSNR values the CSC will check these changes with the network operators and in case of the root cause of these changes are unknown the CSC will send trigger pulses and the initialized keys to activate the security layer in the source and destination stations, and it will define the clock rates of the proposed security model. With every clock pulse, the dual synchronized Random number generator (RNG) will generate two independent outputs. The 1'st output is the generation of one Random polynomial with degree n out of 2^{n-1} Polynomials to enable or disable the switches of the LFSR security model. The 2'nd output is the generation of one initialized secret key which was received before from the CSC with length $l_k = n$ out of 2^n keys to initialize and rest the key of the LFSR model with every frame in the OTN transmission system.

Afterward, the reset cycle starts with the trigger of every clock in both sides of source and destinations and the system makes the encryptions with the same polynomial function and the produced keys from the combinations of the initialized key and the shifted data bits of the original data message with a length $l_m = l_n$.

In case the attacker can manage and access the fiber cable by splitting the optical laser signals and take a live copy of the OTN frames to start the de-mapping process, he will only get the encrypted ODU_k instead of the original ODU_k which included in the OTN frame, as shown in 4.11.

Fig. 4.11. The Structure of the OTN 10/100 Gb/s frame with encrypted ODU

The attacker will need to be aware of 4 variables to understand the encrypted ODU_k, and to be able to retrieve the original ODU_k of the OTN frames

- The 1'st variable is to know the algorithms of the dual SRNG with its synchronization clock.
- The 2'nd variable is to know the secret keyword which used in the reset cycle with every trigger of the clock pulse which has the probability $p_k = 1/2^n$ also, entropy equal $\sum_{i=0}^{n} k_i (\log 1/k_i)$ as listed in equation (4) to initialize the original message.
- The 3'rd variable is to know the polynomial degree, with the probability $p_p = 1/2^{n-1}$ moreover, entropy $\sum_{i=0}^{n-1} P_i (\log 1/P_i)$, as listed in equation 4.13.

Chapter 5
Operational Cost Optimization of the 4G Traffic over the Optical Transport Network

5.1 Introduction

Working in the operational tasks of the optical network by the traditional methods is very expensive and directly affects the net profits of any telecom operator. To optimize the operational costs in an extensive optical network, many tasks that should be done intelligently by using the suitable machine learning technique. The most important factors which affect the cost of the operations in an optical network are how they can monitor and control the energy consumptions, reduce the time of the fault localization, monitor the quality of the transmission links in a periodic way and finally choosing the best routes for the creations of the new circuits to reduce the waste of the rare network resources. The previous studies indicate that the artificial intelligence and the machine learning techniques have the promising future in the automation and the smart applications in many fields such the optical network, as the previous studies investigated separately one factor only from all of the operation factors such that: many studied introduced the AI to predict the energy consumption by using artificial neural network (ANN) and support vector machine (SVM) in the renewal energy field [62]. Other studies explained how to use the different machine learning techniques in the forecasting of energy consumption in the electricity market by using multi-objective genetic algorithms (MOGAs) [63]. Few studies investigated how to optimize energy consumption in the Datacenters by using the AI forward and backpropagation techniques [64]. Many different studies explained how to use machine learning algorithms to predict the fault location in the optical network by using a double-exponential smoothing and support vector machine (DES-SVM) algorithm [65]. Other investigations used the graph-based correlation (GBC) heuristic in defining the fault location in the optical network [66]. A long short-term memory (LSTM) network with the in-depth learning technique was introduced by certain studies as a solution to continually estimate the optical signal to noise ratio (OSNR) [67]. Few previous studies were done about how to use the AI techniques in the configurations management of optical network to automate the creation of the new circuits on the optical network one of these studies explored how to use the dynamic routing and the spectrum assignment with the hold time between the different nodes in the optical network to define the best routes in the Dense Wave Division Multiplexing (DWDM) network [68]. The chapter proposes, for the first time, a universal platform to derive the critical operational factors in the optical network to the automation, and this done by the interactions between four different proposed models to automate the executions of the operational tasks in the network. The Four models of the operational tasks in the optical network are introduced and implemented by the different techniques of the Artificial Neural Network (ANN) to reduce the human interventions in the executions of these factors and at the same to optimize the waste of the resources in the optical network.

5.2. The operations of the optical network

The optical transmission network is considered the core infrastructure of the communication industries. This network was built over many years, and it is still extending overtime in all country areas.

One of the most critical challenges in this network is the enormous diversity in its resources, as an example of this diversity it consists of many different vendors equipment, different transmission technologies from the synchronous digital hierarchy (SDH) to the Optical Transport Network (OTN) with the Core DWDM technology, and different spans in the fiber cables of the network.

As a result of this diversity in such huge amount of optical network resources many network management systems (NMS's) were used to manage it (one NMS for every vendor equipment type), and this makes the executions of the operational tasks by the human interventions between these different NMS's very difficult and expensive, at the same time the energy consumptions over all the network of these NMS's is not controlled by centralized management system. The results of the previous situations are high costs of the operations, tremendous efforts, and resources with a long time to meet the target of the Operations key performance indexes (KPI) over the entire optical network. Figure 5.1 shows an example of the optical network with many vendors and NMS's.

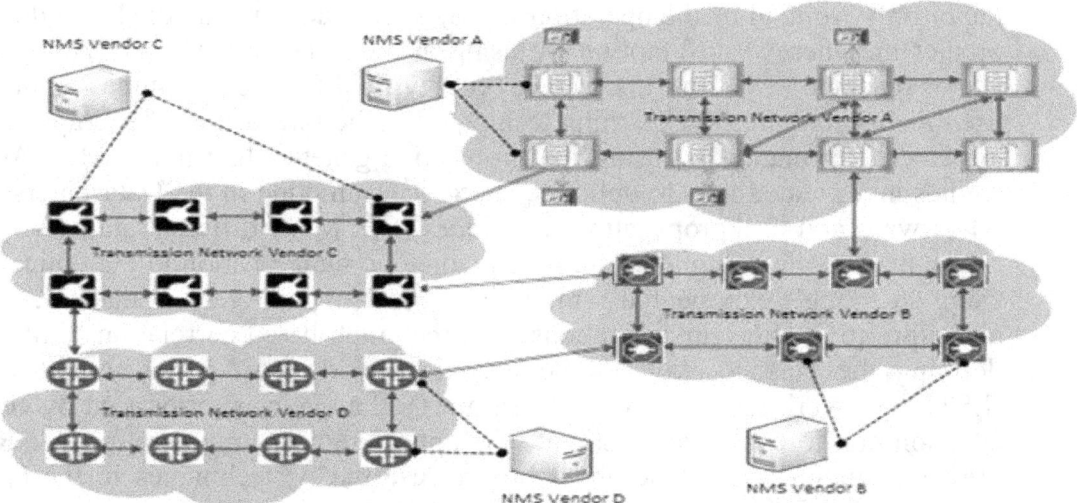

Fig. 5.1. Example of the Transmission Network with Multi-Vendor

As shown in figure 5.1, the executions of the operational tasks in the network of different vendors are very complex. In this paper, we study only four Use-Cases in the operations task of the optical network with different vendors and different NMS's to illustrate four challenges in the network.

5.2.1 Monitor the total Energy consumption over all-optical networks

One of the most critical challenges in the optical network is how to optimize energy consumption, especially with the considerable rise in the cost of providing it from other parties. The AS-IS case in monitoring the energy consumptions is built upon every owner of the network element stations calculates the consumption of the power according to his perceptions and the following equations:

$$P_{DC}(NE) = I * V \quad (5.1)$$
$$P_{AC} = \sqrt{3} I * V * \cos \emptyset \quad (5.2)$$
$$P_{TL} = N * LAMP\ WAT \quad (5.3)$$

Where $P_{DC}(NE)$ is the DC power consumptions for one Network element, P_{AC} is the power consumptions by the cooling systems, P_{TL} is the power consumptions by the lighting system, N is the number of the lighting unit, $\cos \emptyset$ is the Power factor, I is the consumed current, and V is the Power Voltage Value.

As we can see from the previous equations there are no relations between the power consumed by the network elements and the power consumed by the infrastructures such cooling systems and lighting systems, also there are no relations between the estimations of the power consumptions in one station to the remaining part of the optical network which is associated with the loaded services on the optical network elements.

Challenge no.1: There is no centralized system that can be used to monitor and control the performance of the total power consumptions (the NE's consumptions and infrastructure consumptions) and the relations between these consumptions with the total loaded services in the network elements over the entire optical network.

5.2.2 The time of the fault localization

The supervision of the optical network with such complexity and heterogeneity as shown in figure 1 is limited by the manual administrations of the operational tasks, and the probability of the risks in the optical network increases with the complexity of the network and how much the execution of the operational tasks is done by the human interventions. One of the most critical challenges in the supervision of the optical network is how to find the location of the faults and how to easy define the root cause of any affected traffic within a suitable time. In the traditional methods, the fault localization depends only on the observations of the different alarm lists from many NMS's by the responsible persons, which may contain hundreds of records [69]. The time consumed in the fault localization phase and the quality of performing this task depends mainly on the experience of the responsible persons. This method of fault recovery affects the operational cost by increasing the meantime to repair (MTTR) directly and maybe violating the service level agreements (SLA) with the customers.

Challenge 2: There is no automated process to faster the fault localization time and find the root cause problem to minimize the meantime to repair over the entire optical network

5.2.3 The Optical performance monitoring

The quality of the transmission network depends mainly on the proactive maintenance in the network. Monitoring the quality of the transmission links over time is an essential task in the operations of the optical network. One of the critical factors used to measure the performance of the optical network is the optical signal to noise ratio (OSNR) [67]. The traditional method in monitoring the optical signal to noise ratio is done by the observations of many lists, which contain thousands of laser power measurements in the different NMS's systems. It takes a long time and considerable effort to keep in track with the dynamic changes in the power measurements in all the optical ports of the network. The quality of performing this task depends only on the experiences and the commitment of the responsible persons. Any improvement in performing this task will affect the number of complaints from the bad quality of the network and affects the operational cost of the optical network positively.

Challenge 3: There is no automated process to perform the performance monitoring task and notifies by the needed proactive actions in the optical network

5.2.4 Configuration management

One of the most critical issues in the optical network is the utilization of the optical network. To create a new circuit in a multivendor network with the traditional way it will take a long time to complete the route creation.

The quality of the network configuration is varied according to the performance of the creations and the existing gap between the high level of the requirements and the low level of the configuration methods [69]. The quality of the network configurations affects the operational costs in a negative way such that in case of bad quality will make waste of the expensive resources. The time consumed in different steps in the new creation process in the optical network depends on the type of customer request. The time consumed in the implementation of the customer request in the network is the longest in all creation period. In case the creation is done over a multi-vendor network the time, and the efforts will be doubled according to the number of the vendors in the network.

Challenge 3: There are no automated tools to perform complete provisioning of all resources in the optical network and to maximize the return of the investments in these resources.

5.3 The Proposed Solution for Smart Operations in the Optical Network

From the previous investigation, we found that artificial inelegance is the best solution to perform a smart optical network. The proposed solution to overcome the previous challenges in the optical network is to design an Intelligent Universal Platform (IUP) in the upper layer than the NMS layer.

The proposed platform is done to transform the operational tasks in the optical network to be more automated and is working as robots instead of human interventions. One of the essential functions of the proposed platform is to predict the optimal solution to execute the operational tasks. The IUP consists of the interactions between 4 proposed models to perform the optional tasks in the optical network.

Every model is designed by using machine learning techniques or the deep learning techniques to reduce the human interventions in the processes of the operations, especially in monitoring and predicting the optimal solutions for performing the operational tasks. The design of every model and the complete solution model are explained in the following subsections.

5.3.1 The Power Consumption Model

The model consists of a machine learning algorithm that learns from the actual power consumption in some sites in the network. The training data of the system is raised from m optimal sites as following equations [70, 71]:

$$PUE = \frac{Total\ Facilities\ Power}{Optical\ Network\ Equipment\ Power} = \frac{FP}{EP} = \frac{F_0 + \sum_{j=1}^{n}(a_j) + \sum_{p=1}^{c}(b_p)}{\sum_{l=1}^{p}(E_l) + \sum_{i=1}^{k} w_i(m_i + s_i)} \quad (5.4)$$

Where PUE is the Power usage effectiveness in the Optical Network, F_0 is the initial power consumption for the facilities in one site, a_j is the power consumption by the cooling system with an air conditions in one site, b_p is the power consumptions by the other facilities for c units (lighting – measurement equipment – fans), E_l is the power consumptions by p empty optical network shelves, m_i is the power consumptions by inserted i modules in the total number of shelves, s_i is the power consumptions by the implemented services on the i modules, and w_i is the weight of the power consumption for i modules with its services.

The proposed model of power consumption consists of 3 layers of the Artificial Neural Network (ANN), which are used to form a simple prediction model of the power consumption performance in the optical network, as shown in figure 5.2 [72].

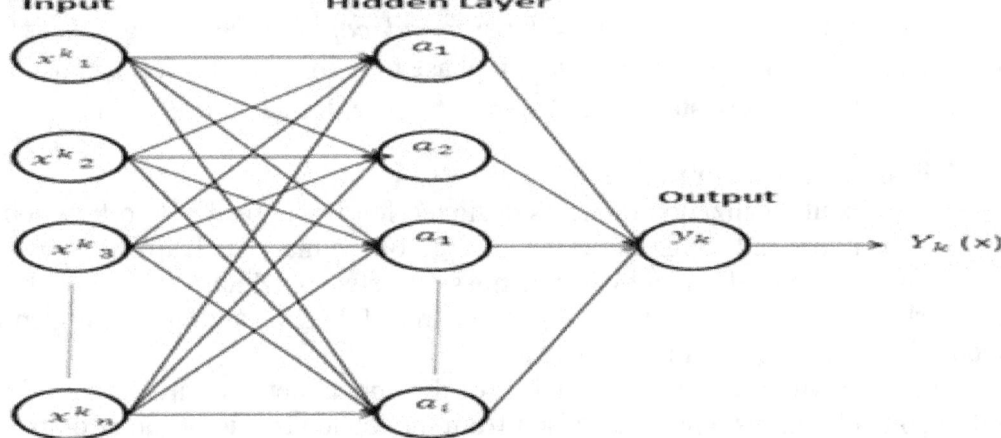

Fig. 5.2. The Proposed Artificial Neural Network for power consumption model

- The first layer is assigned for the input variables x^k_n Of the different parameters in the optical network such as cooling, air-condition, lighting, modules, number of the network elements in the same network elements, and any other variables with n number of variables. k number of training examples
- The 2'nd layer is the hidden layer which used to find the relations between the output and the input variables with a total number of i coefficients a_i

- The 3'rd layer is the output layer, which explores the PUE value of the optical network in one site.

The deployment of the model is done in 3 steps:
- The learning phase: where the output $Y_k(x)$ moreover, the input x^k_n are known from the training data with the following equation, which explains the nonlinear relationship between the outputs and the inputs [71, 72].

$$Y_k = Y(x^k_n; a_i) = \sum_{i=1}^{i=h} a_i f\left(\sum_{n=1}^{n=m} a_i x^k_n\right) \quad (5.5)$$

Where k expresses of the site number during the training phase and a_i is estimated coefficient function to perform the relation between inputs and outputs

- The generalization phase in this phase is evaluating the machine learning algorithm to generalize it on all network by testing the expected outputs $\widehat{Y_{\bar{k}}}$ in site $\bar{k}$ with the unseen examples in the training phase as the following equations:

$$\widehat{Y_{\bar{k}}} = \sum_{i=1}^{i=h} a_i f\left(\sum_{n=1}^{n=m} a_i x^{\bar{k}}_n\right) \quad (5.6)$$

To evaluate the algorithms, the root means square errors (RMSE) will be used as the following:

$$RMSE = \sqrt{\frac{1}{N} \sum_{\bar{k}=1}^{\bar{k}=N} (\widehat{Y_{\bar{k}}} - Y_{\bar{k}})^2} \quad (5.7)$$

Where Y_k is the actual output of the learning data set, x^k_n is the n input variables for k training samples, a_i is the coefficient functions of every neural cell, $f(x)$ is the function of the hidden units are related to the tangent function such $(e^x - e^{-x})/(e^x + e^{-x})$, and $\hat{y}$ is the expected output of the testing data set.

- The last phase in the implementation phase to monitor and test the consumption power over all the network stations and alert with the extreme values in the consumptions.

5.3.2 Fault localization model

The Fault localization model is designed for any optical network which includes different types of vendors and technologies, the design of the machine learning model is done according to the assumptions of the optical transmission network which consists of the core transmission network with dense wave division multiplexing (DWDM) and optical transport network (OTN) technologies as shown in figure 5.1.

As shown in figure 5.3, the faults of the optical network are classified into two categories the secondary alarms which are raised from the loaded traffic of the circuits on the network and the primary alarms which are raised from the modules and cards of network elements [73]. To formulate the training data of the machine learning algorithm, many correlation rules between all the alarm types should be formulated first.

The correlation rules are built according to the time factor as the secondary and primary alarms are raised to the Network Management System (NMS) at nearly the same time, these rules are depended on the mean time between consecutive incidents (MTBCI) which is very small for the same root cause incident in the optical network, by considering the primary alarms which express the network alarms is time-independent and the secondary alarms are time-dependent.

According to the Weibull distribution, the localization of the failure links depends on the correlations between the different variables of the alarm lists, which stored in the change management database (CMDB) the different NMS's.

The data set used in the learning model was created from the past practical experience according to the following techniques [66, 74]:

- The affected equipment raises abnormal conditions to the NMS with many records in the list of alarms A(i) which consist of two groups of data secondary alarms S(i) and Primary alarms P(i) where A (i) = S(i) + P(i) as shown in figure 3. The secondary alarms are associated with certain links in the optical networks by the Vector failure links X (i) = [X_1 (i), X_1 (i) ... X_D (i)] and can be determined as following [66]:

$$X_j(i) = \begin{cases} \frac{S_j(i)}{A(i)}, & D \geq j \geq 1, X_j \in S(i) \\ 0, & otherwise \end{cases} \quad (5.8)$$

Where X (i) is the distribution vector of the failure links from a total number of links in the optical network L.

Fig. 5.3. The Types of the data in the alarm list of the NMS

- The root cause of the faults incidents is the link failure in the optical network which is one link from the Vector X (i), and determined as the following equation:

$$Y_j(i) = \frac{P_j(i)}{X_j(i)}, \quad X_j(i) \subset X(i) \quad (5.9)$$

- If Y_k (i) is equal to any positive value, then the link with name k is the root cause of the total failure in incident i. The implementation algorithm in machine learning consists of 3 layers. As shown in figure 5.4; the first layer is the input of all alarm lists; the 2'nd and 3'rd layers are the definitions of the relations between the failure links, and the alarm lists the 4'th is the outputs of the failed link. The prediction algorithms are implemented according to the nonlinear regressions between inputs and output as the following equation:

$$Y_k = Y(x^k_n; a_i; b_j) = \sum_{i=1 \& j=1}^{i=h \& j=d} a_i b_j f(\sum_{n=1}^{n=m} a_i b_j x^k_n) \quad (5.10)$$

The root mean square errors (RMSE) is used to test the validity of the system by calculating the errors between the actual results of the model Y_k also, the result of the unused test examples in the training phase.

5.3.3 The optical network performance model

The model is as intelligent optical performance monitoring (IOPM); it covers all the OSNR Monitoring tasks for the high rate of the wavelengths in the multi-vendor DWDM network. Also, it supervises the end to end optical layer performance in the whole optical network. The training data is used from the change management database (CMDB) of the NMS systems and are classified according to every Optical port type in the NE as following: Transmit power, receive power, Error rates, Receive End of Live power (EOL), Optical Fiber span attenuation and ONSR. Table 5.1 illustrates the different types of data, which are used as input or output in the IOPM model for Every NE [75, 76].

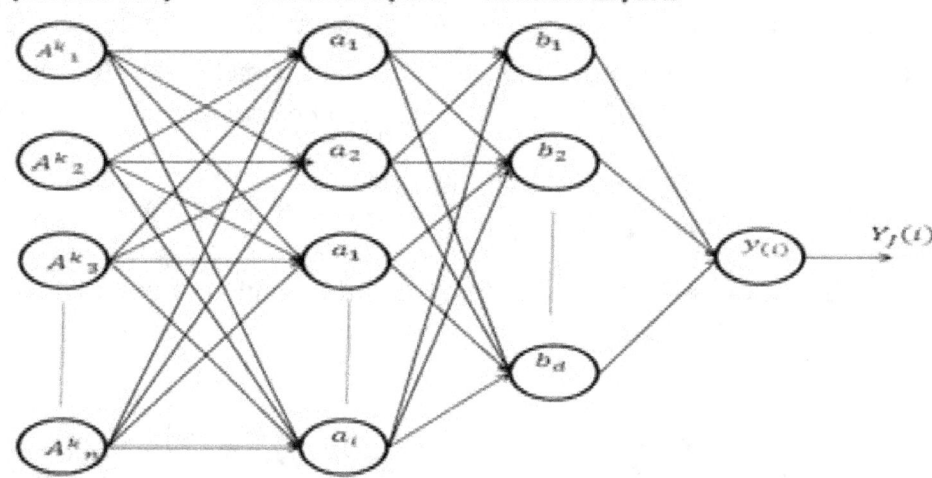

Fig. 5.4. The Proposed Artificial Neural Network for fault localization model

Table 5.1. The Variables of IOPM.

Parameter	Transmitter port	Receiver Port	Amplifier card
Optical power	Input	Input	input
BER		Output	
OSNR	Output	Output	output
Amplifier Gain			input
Fiber Span Attenuations		Input	
Line Rate	Input	Input	

The first phase of building the model is the learning phase, and the data set in this phase is obtained from the CMDB of the NMS for every NE with optical ports bit rate 10 Gb/s or more. There are linear relations between output factors (OSNR & BER) and the input factors as following [77]:

$$OSNR = 10\frac{S}{N} \quad (5.11)$$

$$10\ log_{10}(BER) = 10.7 - 1.45\ (OSNR) \quad (5.12)$$

Where S represents the linear optical signal power, N represents the linear optical noise power.

The model consists of 3 layers one layer for input parameters, which consists of n variables and d ports, and the 2'nd layer from the relation between input and output with m coefficient, and the third layer is the output layer, as shown in figure 5.5 [4, 78].

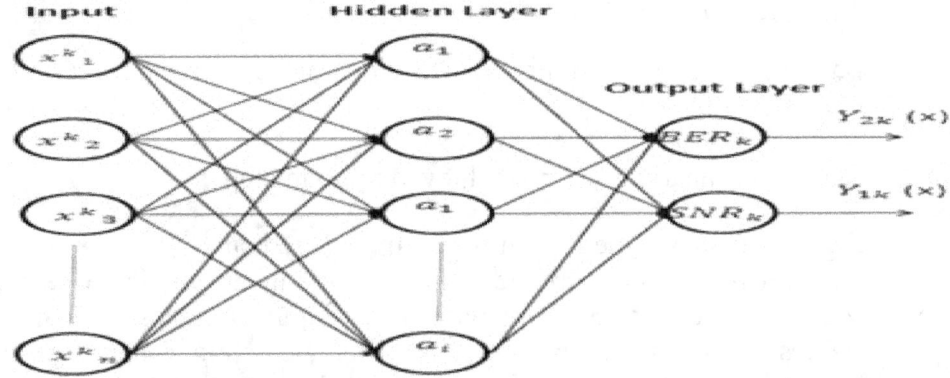

Fig. 5.5. The Proposed Artificial Neural Network for Optical Performance model

$$Y_{1k}(x) = f(\sum_{n=1}^{n=d} \sum_{i=1}^{i=m} a_i x_n^k + w_0) \qquad (5.13)$$

$$Y_{2k}(x) = f(\sum_{n=1}^{n=d} \sum_{i=1}^{i=m} a_i x_n^k + w_1) \qquad (5.14)$$

Where $Y_{1k}(x)$ is the expected output of the OSNR in the port number k, $Y_{2k}(x)$ is the expected output of the BER in the port number k, a_i is The coefficient functions of the hidden layers, f(x) is Each of the hidden units is related to the tangent function, x_n^k is the input of n variables for port number k in the network, w_0 is The fixed weight between input variables and output OSNR, and w_1 is the fixed weight between the input variables and the output BER.

5.3.4 The configuration model

In the configuration model a combinations between the Artificial Bee Colony Algorithm (ABC) and the ANN is used where the model consists of Artificial Neural Network (ANN) which uses input topology, physical layer characteristics, Optical port status, capacity, source and distention to be implemented according to the ABC technique in 5 steps [79]:

- **Step 1** generates a list of paths randomly between every node as a source and all possible destinations by the process of the search paths and the spanning tree control to form the routing table for all networks and update the external archive in the database of the model.

- **Step 2** produces a new solution for every created route in step1 by using the Greedy Selection Process (GEP).

- **Step 3** Search for the new multicast tree from the source node to the destination nodes.

- **Step 4** Choose the best probability for the routes from the sources and destinations according to the one with smaller D value has a higher choosing probability as the following equations [79, 80]:

$$p(\theta_i) = \frac{1}{R(\theta_i) + D(\theta_i)} \; ; \quad \theta_i \in S \qquad (5.15)$$

$$D(\theta_i) = \frac{1}{2 + \sum_{k=1}^{\sqrt{|S \cup Arch|}} \|f(\theta_i) - f(a_k)\|} \; ; \quad a_k \in S \cup Arch \qquad (5.16)$$

$$R(\theta_i) = \sum_{\beta < \theta_i} \{w | \beta \prec w \wedge w \in S\} \qquad (5.17)$$

Where $p(\theta_i)$ is the chosen probability for the route θ_i, S is the set of the routes from one source to the possible destinations, $D(\theta_i)$ is the candidate's density around θ_i based on $\sqrt{|S \cup Arch|}$ nearest neighbors, and $R(\theta_i)$ is the importance of θ_i terms of how many candidates are dominated.

- Step5 Produce new alternatives according to the current probability and update the archive data.

- Step 6 the terminations are not achieved to go to step 2.

Figure 5.6 shows the design of the model by using four layers of the ANN model. The first layer is the input variables which are the inventory database from the CMDB of the NMS of the optical network with many input variables such number of available ports, nodes, capacity of every optical link, existing routes, available frequencies, topologies of networks, types of every node, the sources, destinations and the cost of every link. The 2'nd two layers are the hidden layers, which will be used to find the relations between the input and the output; the final layer is the output layer. The implementation of the model is the same as equation 5.10.

5.3.5 The Total Proposed Platform

From the previous subsections, we found that there are direct relations between the different variables of every model with the other. Table 5.2 illustrates the relations between the inputs variables and the output variables for all models in one site of the optical network, as all variable is extracted from the CMDB of the different NMS's.

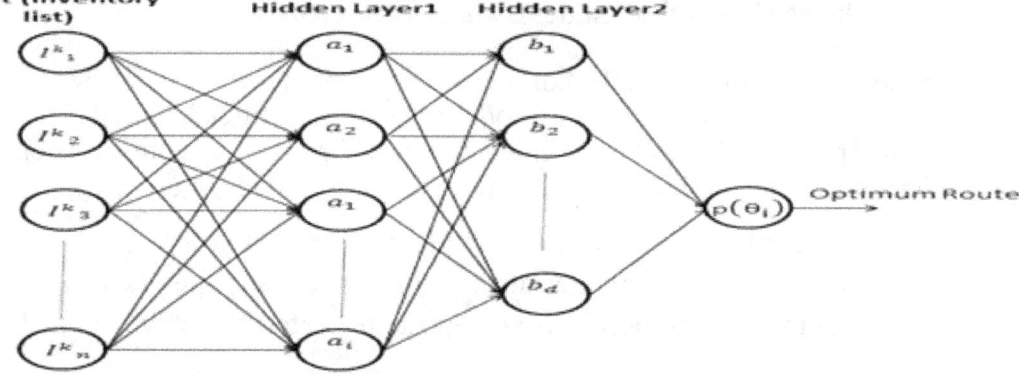

Fig. 5.6. The Proposed Artificial Neural Network for the configuration model

Table 5.2. Total Variables of the universal platform

Input Variables		Output Variables			
		PUE	Fault Location	OSNR Performance	Optimum Route
Number of NE's	NE	■			
Traffic weight	TW				■
Power consumed for one NE	PN	■			
Power Consumed for cooling system	PC	■			

Power Consumed for other infrastructure systems	PI					
Capacities of every optical link	CO					
Transmit Laser Optical power for every port	TL					
Receive Laser Optical power for every port	RL					
Bit Error Rate for every link	BER					
ONSR for every link	OSNR					
Secondary Alarm lists	SA					
Primary Alarm Lists	PA					
Number of the optical links	NL					
The topology of the network	TN					
The cost factor of optical links	CF					
Source	S					
Distention	D					
NE Vendor Type	NV					

Figure 5.7 illustrates the proposed complete model by using the ANN technique to execute the operational tasks in the optical network. The general model consists of 6 layers the first layer for the input variables as shown in table 5.2, and source of the input data set comes from the CMDB of the multi-vendor NMS's, the 2'nd layer is the first hidden layer and consists of m neural cells which are connected with all the input variables, the 3'rd layer is hidden layer with k neural cells, and it sorts the input variables according to the output types, the 4'th layer is hidden layer with n neural cells, and it makes more classifications to the variables according to the output types, the 5'th layer is the last hidden layer and consists of q neural cells, and it determines the different forms of the outputs, and finally the last layer is the outputs. The general model is represented by the following equation [81, 82] :

$$Y_k = Y(x^k_n ; w_f ; a_i ; b_j ; c_l ; d_r) = \sum_{f=1}^{m} \sum_{i=1}^{k} \sum_{j=1}^{n} \sum_{r=1}^{z} \sum_{l=1}^{q} w_f\, a_i\, b_j\, c_l\, d_r\, f\left(\sum_{n=1}^{n=m} a_i\, b_j\, x^k_n\right)$$
(5.18)

Where Y_k the chosen output of the model is, x^k_n is the data set of the inputs, and $w_f ; a_i ; b_j ; c_l ; d_r$ is the coefficient functions of the hidden layer respectably.

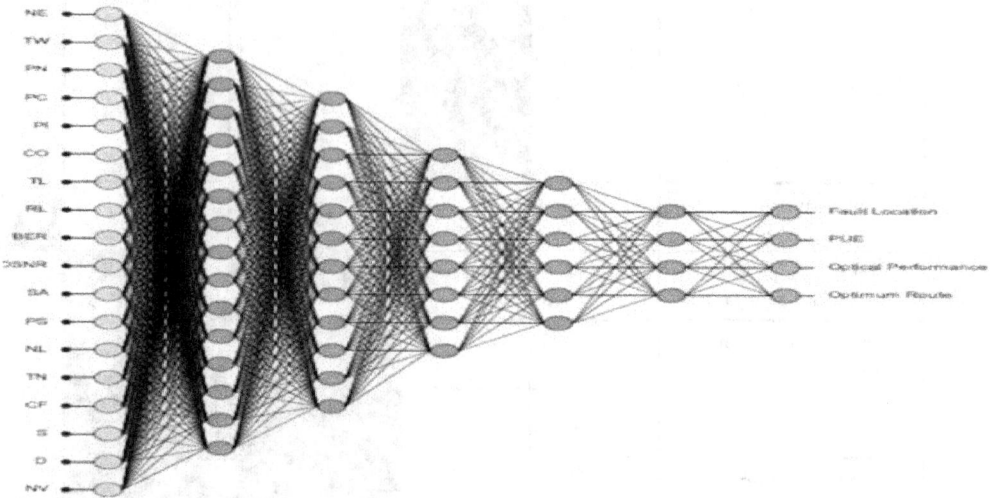

Fig. 7. The Proposed Artificial Neural Network for the General Model

5.4 Conclusion

For the first time, we propose an Intelligent Universal Platform (IUF) to manage and optimize the operational tasks in the optical network. The proposed platform drives the supervision of the network to the automation, and is formed by the integrations between four models, the 1'st one is the energy model and built according to the variations in the power consumptions and it proposes the recommended actions to optimize the energy consumption by swapping the cooling system between two system types (air-conditions and fan units) according to the temperature degrees , also the model learn from the variation in the heat dissipation according to the busy and the idle modules and it can transform the unused hardware to the sleeping mode, the 2'nd model is the fault localization and is built by the practical experiments of the faults recovery and forms its expert system to predict the meantime to repair (MTTR), the 3'rd model is the network performance and is built by monitoring the optical signal to noise ratio (OSNR) and provides the recommended corrective actions to keep the same level of the performance, the 4'th model is the configuration management, and it suggests the best routes in the network. The paper studied four use cases in two situations As-Is and To-Be, the cases are about the energy consumptions, the fault locations, the circuit creations, and finally, the variation in the signal to noise ratio. The results show that by using the machine learning in our platform the time of the fault location is reduced from 40 min to 3 min, the efforts to create one circuit is reduced by 30.87%, the number of the complaints are reduced by 30% per year, and the response time to the number of the complaints is decreased from 55 min to 5 min, This indicates that shortly the machine techniques will play a significant role in monitor, detect, localize faults, and finally optimize the resources of the optical network, all without human intervention.

Chapter 6
The intelligent Relicense of 4G Traffic over the Optical Transport Network

6.1 Introduction

At present and soon, the needs to exchange a huge amount of data between the different regions inside the country become a particular commitment especially with the significant expansions of the clouding principles, the data centers and the implementations of many intelligent applications of the 5G in the several fields of the information technologies [1]. The existing model of the core transmission network consists of 3 autonomous zones, the 1'st zone is layer 1 and layer 2 which considered as the physical layer of the backbone optical network such OTN, the 2'nd zone is layer 3 which considered as IP core network, and the last zone is the access and the application layers. The challenge in this model is that; there are no correlations between these 3 zones of the communication network to optimize the capacities and the demanded resources which are needed to transport the data between each of them. Nowadays, the optical transport network (OTN) represents the crucial position of carrying a tremendous amount of data between the different sites according to the demands of the consumers of telecom operators. The structure of the optical transport network (OTN) in our case is a practical case study of the long distances optical network in one of the middle east countries, the model consists of multiple domains of transmission equipment's from different vendors, and connected with each other's by thousands of kilometers from the fiber cables, at the same time every area inside the country is covered by 1 or 2 isolated domains from the long-distance OTN with the same or different vendor types. The problem in this traditional structure of the long-distance optical transport network is the moderate quality of the services resilience and the challenge to reuse the free bandwidths on a certain optical area domain, this is a consequence of the existing optical network was built from thousands of new and legacy network equipment's (NE's) with many different types, capacities, and vendors in long intervals of times. This style of the building the optical network converted it to isolated domains of the optical network with chaotic merging between the legacy layers such as the synchronize digital hierarchy (SDH) and the new generations layers as the OTN equipment. Although there are many efforts were done from many international organizations to create standardizations among the vendors in the optical network fields to shape their network parameters under the same umbrella of the regularities, every vendor has his supervision system, provisioning standard for the operation, and the administrations of his optical network [83].

With the existing structure of the OTN, it is challenging to restore the traffic from certain area domain to other area domain in case of any crisis will be happening in each of them, especially with different vendors for each area even though these area domains are joined in common get-way stations. Moreover, the routing between the different OTN domains needs more manual and hardware arrangements to implement it, which is not proper during any disaster in the core OTN and the critical needs to return the affected services.

The new generations of the communication businesses for the next 3 years will involve the clouding concepts in the various fields of the communication markets, and will expand in the implementations of many intelligent applications following the employing of the 5G technologies around the world.

All of these tremendous advances in the communication markets in the near future will need exceptional demands in the infrastructure of the core optical network, which depends essentially on the performance, the latency, the available bandwidth, the dynamically of the recovery of the service from certain paths to others, the switching rate and other parameters in the OTN. To fulfill the conditions of the new generations of the communication markets, the current OTN should be reconstructed from a Rigid network to be more dynamic and smart network [84]. Transforming the OTN to be more dynamic will support the operators to extract the most regular routes in the overall network and will enhance the availability to restore any affected traffic in the shortest time, even though among the various vendors and the complex area domains [85]. In this chapter, for the first time proposes an Intelligent Software-Defined Optical Network (ISDON) by offering the alliances among the software-defined network (SDN), the machine learning technology (ML), and the traditional network management systems (NMS's) of the OTN, This will be done to convert the present rigid OTN to be more dynamic and automatic, and this will be done applying just one centralized controller (CC) to maintain and master the OTN in the various vendors' fields, different layers, and in the domain of the different areas. The CC will make its choices about the restoration routes, the switching time, the latency, and the other network parameters according to the past experiences of the network, which was built by the machine learning (ML) model. Also, the concept of the optical network cloud is offered between the different network areas to deliver the traffic restorations between the different vendors' equipment in the IP and the OTN domains [86, 87].

6.2 Resilience Challenges and Motivations

Figure 6.1 demonstrates a practical model of the long-distance optical transport network in this paper, the model expresses a small representation of a real long-distance transmission network in one of the Middle East countries, and consists from several sectors of the optical transport network, where every sector called domain, and every domain covers large area inside the country, and consists of DWDM equipment and OTN equipment from various or the same vendors, the capacities of every domain reaches to 8 Tb/s on the line bandwidths with 100 Gb/s for each wavelength bandwidth in the domain. In this paper we concentrate on the challenges of the resilience and the routing mechanisms between the different domains in the long-distance transmission network and the role of the SDN with the ML to enhance the stability of the services flapping and the routing mechanisms between the different domains of the legacy optical transport network especially in case of the multi failure or any crisis in one or more domain.

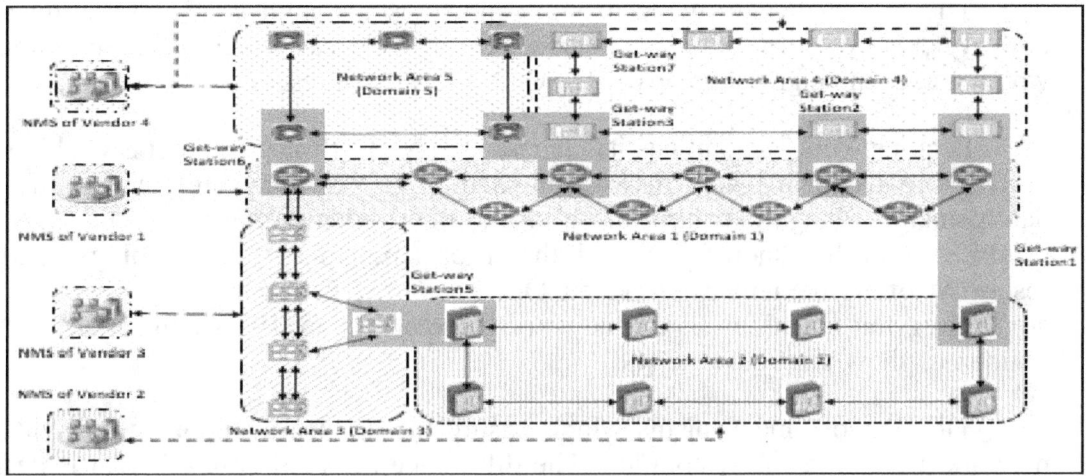

Fig.6.1. Practical Model of the Long Distance Optical Transport Network

According to the current OTN long-distance model as shown in figure 1, there are some constraints which affect the resilience flexibility and the performance of the network, such the few number of the possible routes in the optical fibers cables, which depend on the characteristics of the geographic area inside the country, particularly in the middle east countries which have large area of deserts and mountains with very long distance between the consecutive network elements. As a result of these limitations, the traffic in the long-distance optical network which connects the essential cites with each other's is protected by two routes only, known as working and protection routes; seldom there is 3'rd route or more for restoring the traffic during the multi-failure within the same domain. Other restrictions in the current model is, there is no any integrations or alliances between the different domains of the OTN, which were developed before based on the vendor type, every network area is separated in the topology, the associations between the working and protection routes with the corresponding domain only, one network managing system (NMS) for every vendor type which is network for its domain only, and the available bandwidths are kept for its domain area only. Other difficulties of this model is that although there are many stations are common with more than one network area or domain, it's tough to provide backup routes to the working services between the different areas domains in case of any failure or crisis in one of these domains, in other words it will take a long time to prepare manual restoration of any impacted traffic from one area through other area domain by using the common get-way station among each of them even though the same NMS or the same equipment type between the different network area domains. Other challenge in the existing long-distance OTN model is that, there is no any tools in the OTN network to take the performance and the delay of the backup routes in the consideration before choosing it for the restoration or to keep the services on the routes which have lowest switching rates or lower number of failures, although the network is switched many times within short period as a result of many cable cuts in the long distances routes, at the same time most of the long-distance cables had high attenuations due to the multiplied cuts on it, in other words, the control planes of the resilience's mechanisms in the different area network has no techniques to select the best backup routes in the performance and the latency according to the experience of the long-distance optical network. Another challenge in this model is that it is challenging and need a long time to reuse the free capacities in a certain domain to cover other network areas in case of any congestion in this area without equipping the connections between the get-ways in an old-fashioned way.

The last challenge in the existing model is the high cost of building a long-distance optical network with low revenue, especially in the areas which have low population density, which makes the high capacities of the OTN are wasted resources.

Each network element has its Generalized Multi-Protocol Label Switching (GMPLS), the GMPLS controller of every network element (NE) communicates with each other's over the data plane, by using the control plane to find the available routes over the OTN network. Each GMPLS controller should have all the information about the topology, and the available resources of its domain network, an Open Shortest Path First (OSPF) protocol is used for advertising the situations of the network between the GMPLS controllers of NE's inside its domain as shown in figure 6.2 [88].

The network mechanisms which ensure the service availability is divided between the network protection and restoration. The difference between both of them is defined according to the allocation of the network resource before or after the occurrence of the network failure. There are many different types of the restoration, and the protection mechanisms inside every network domain, one of these mechanisms is called 1+1 protection, where the essential route which is carrying the services between the sources and the destination is called working path, and the other route which is reserved for carrying the services during the failure is called backup route [89].

This scheme of protection mechanism provisions the backup routes individually in case of any failure in the network rather than managing the overall routes of the network as one unit to provide the best backup paths for the network failure, which is not optimized scheme. Where in the protection mechanism the resources for the recovery paths is reserved all the time before any failure occurring in the network such as 1+1 or 1:1 protection types, and in this case the services are recovered very fast generally in less than 50 ms, however it requires signaling protocol and synchronization between the 2 failure nodes.

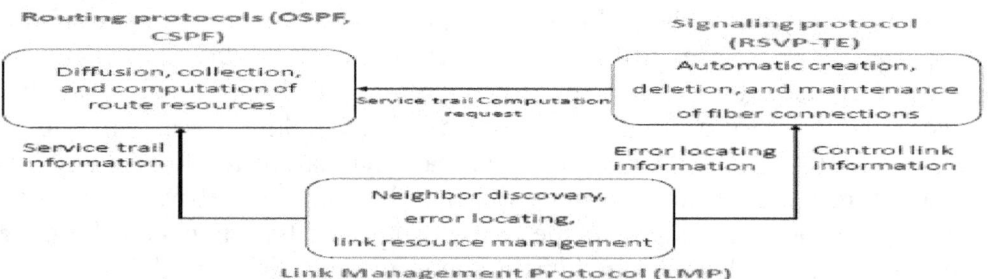

Fig. 6.2. The relation between the ASON-Related Protocols

In the restoration mechanisms, the allocation of the recovery resources is reserved after the occurrence of network failure according to the priority of the affected services, and the recovery processes of the affected services take a long time for almost few seconds to finish the full restoration of the traffic [90]. The automatically switched optical network (ASON) is the mechanism that restores the service in the OTN by using dynamic connection capability, and this technique is done by performing the call and connections control functions through the control plane. To enable the ASON mechanism in the OTN, it contains discovery functions, routing functions, and signaling functions as the following restoration steps [91]:

- In the discovery functions there are several types of discoveries; Neighbor discover its necessary to detect the status of the local links between the adjacencies NE's without using it, and it can be required to configure the status of these links manually at the 3 levels of discoveries, the physical media adjacencies discovery which used to confirm the connectivity between the adjacent ports, the layer adjacencies discovery which is used to confirm the associations between the physical connections and the logical links, and the control entities discovery confirms the association between the control plane and the transport plane of the NE's. The Resources discovery, which determines the available resources, the topology of the network and detects the mismatch in the configuration of the network resources, and finally, the services discovery, which verifies and exchanges the grade of the class of the services in the different administrative domains [92].
- In the routing functions, the architectures and the requirements for the routing in the ASON mechanisms are described in the ITU-T Recommendation G.7715/Y.1706, which includes path selections, routing resources, and the routing diagram.
- The signaling functions are done according to G. 807, where the signaling protocol is essential for fast provision and recovery of the paths after failures, it is necessary for release, create, restore and maintain the connections

The Quality of Resilience (QoR) in the OTN is used to measure the effectiveness of the different types of service recoveries during the network failure according to the following equations [93].

$$F(x) = P_r(T<x) = \sum_{t_i} P_r(t_{i-1}<T<t_i) \qquad (6.1)$$

$$P_0 = P_r(T=0) = F(0) \qquad (6.2)$$

$$A = \sum_{t_i \leq \alpha} P_r(t_{i-1}<T<t_i) = F(\alpha) \qquad (6.3)$$

$$MTTR = E[T] = \frac{\sum_i T \times P_r(t_{i-1}<T<t_i)}{U} \approx MDT \qquad (6.4)$$

$$MTTF \approx MUT = \frac{A}{U} \times MDT \qquad (6.5)$$

Where: $F(x)$ is the distribution functions of the infinite downtime T, $P_r(T<x)$ the probability of having service interruptions at most x with t_i reparations i times, P_0 is the probability of having no services down, A Services exceeds the probability services down exceeds the threshold down, MTTR is the meantime to repair, and MTTF is the meantime to fail.

With The pervious challenges in the current long-distance optical network model it will be not accepted from the customers to have services down as a result of the inability to restore the traffic between the different domains particularly with several applications which depend on the availability of the core optical network and with the presence of the service level agreement (SLA) between the customers and the operators it will be costly for the operators to compensate their customers about any interruption doesn't meet the service availability percentage in the SLA, which can be reached to in some cases around 99.999%.

Although there are many challenges in the exiting model of the long-distance optical network, which affects the services flexibility and the automation of the routing mechanism, there are also many chances to improve the current state to be smarter and dynamic.

One of these opportunities is the dynamic property of the ASON mechanism, which can be worked between the different vendors and the various network area domains if it is controlled by only one centralized controller and one control plane for all the optical network. Also other motivation in the current state of the optical network is the new technology of the OTN is extending all over the country with huge capacities, where every wavelength reached to 100G at least, which makes the availability to build one unified virtual OTN network with high capacity he wavelength be more available and more comfortable between the different network areas. More motivations is using the software-defined network (SDN) in the optical network to provide the suitable tools to separate the control plane than the data plan and provides only one controller for all the network management, which makes its more comfortable now to manage all of these different area networks with multi-vendors by using centralized control plane and only one controller. One more motivation to improve stability the current status of the long-distance optical network is the ability to learn from the past failure experiences in the multiple domains in the OTN by using the ML technology and support the centralized controller of the SDN by the resolutions about the stability of every optical link in the OTN to choose the best stable routes in the network, and to carry the platinum services high rates of services quality and resilience [94].

5.3 Proposed Resilience Model in the Long Distance OTN

The suggested model in the long distances optical network to develop the reliance mechanisms and provide effective bandwidths is formulated upon multi-layers smart and dynamics virtual optical network, this don by adjusting the long-distance optical network in 3 dimensions as following:

5.3.1 Developing the Physical Infrastructure of the Optical Network

This is done by building optical clouds in different layers of the core optical network to provide on-demand bandwidths and the backup routes in case of multi-failures or crisis in any domain of the optical networks, the optical cloud slices the long-distance optical network into 3 layers, the 1'st layer is the core OTN which will be used to connect the different domains of area optical networks by each other's through the get-ways stations, and by using DWDM connections with high the capacities of the wavelengths over all the long-distance areas network. These connections will form the Virtual Cloud Optical Transport Network (VCOTN) on the physical infrastructure of the OTN, as shown in figure 6.3. The objective of this layer is to make the OTN more dynamics between the different areas and between the different vendors and to provide the restoration routes with the required bandwidths between the domains for the high capacity's services.

Figure 6.3 shows the proposed layer1 VCOTN of the long-distance optical transport network. The 2'nd layer of the proposed model is the Services Optical Cloud (SOC), in this layer a unified OTN network is built between the get-ways stations by connecting every get-way station with each other all over the country with reserving over the various domains, the aim of this layer is to provide the bandwidth-on-demand in the fastest time and support by backup routes for the low capacities services.

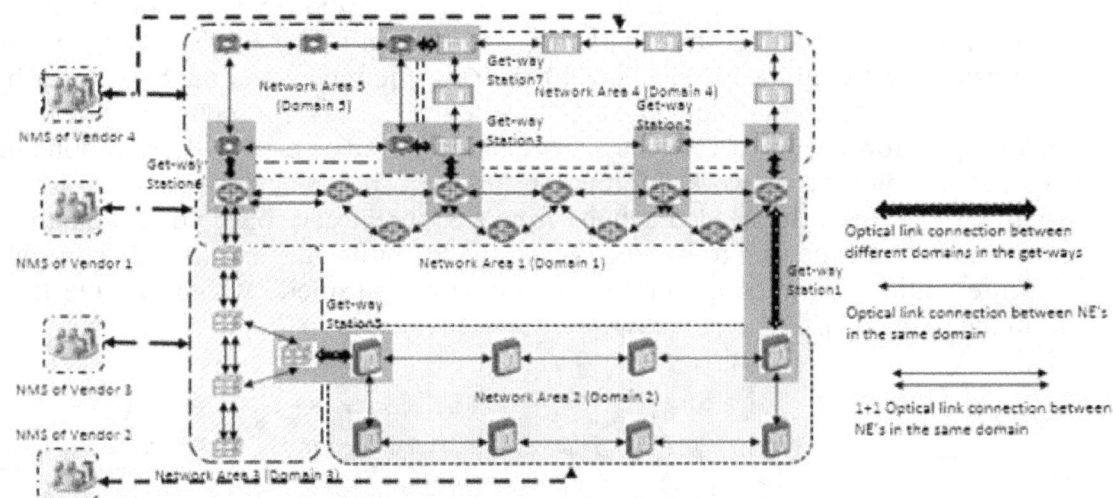

Fig. 6.3. The Virtual Cloud Optical Transport Network (VCOTN) on the OTN

Figure 6.4 shows the SOC in the long-distance network. The last layer of the core optical network is the optical IP core cloud (ICC) which will be built over layer 2 and in the get-ways stations, the aim of this layer is to provide the shortest paths for the restoration routes for IP services and make the services more dynamics with the many alternative's routes over all the core optical network.

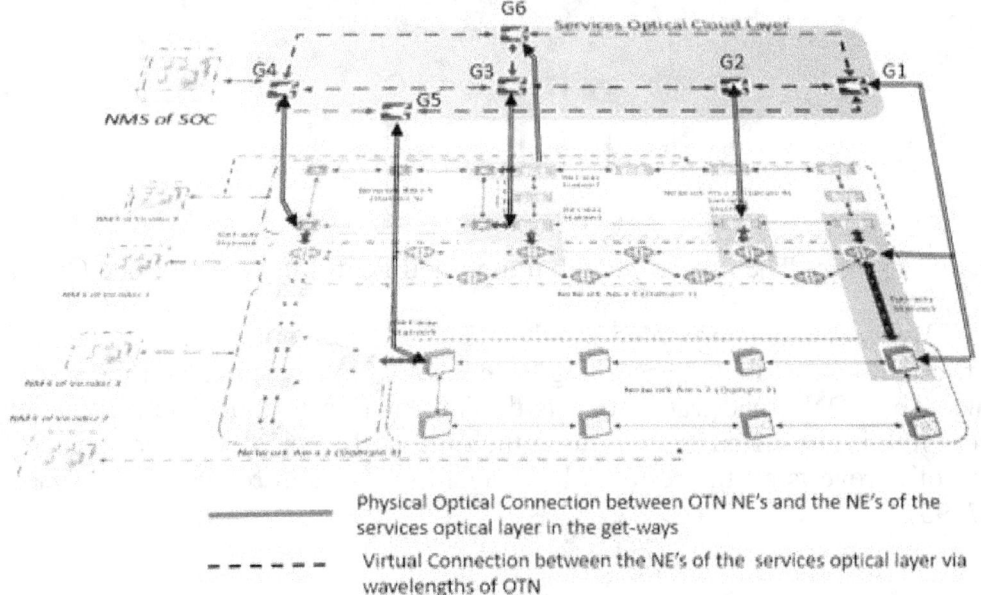

Fig. 6.4. The Services Optical Cloud (SOC) on the OTN

5.3.2 Building the SDN Orchestration

As our transmission model is heterogeneous network and consists of multiple transport technologies, multiple domains with more than 3 optical network layers and many independent network management systems (NMS) to manage the domains of the network, SDN Hierarchical Multi-domain Controller is used to integrate between the different network domains and different NMS's of the optical network. The controller concentrates on the control techniques among the multiple domains, and The SDN orchestration consists of the parent controller, Domain controllers, and abstraction layer, as shown in figure 6.5. The abstraction layer is used to transform between the physical layer of legacy equipment and the SDN layer by virtualizing the network resources. The Parent Controller (PC) has the complete view of the overall network, performs end to end paths computations between the different domains, shares the instructions, and the network information between the other domain controllers. The domain controllers perform the needed actions from the parent controller on its domain only. When a new connection is needed to be established, the PC builds the end to end connections and sends the information of the connection to the domain controllers, which execute the needed parts form them only [5].

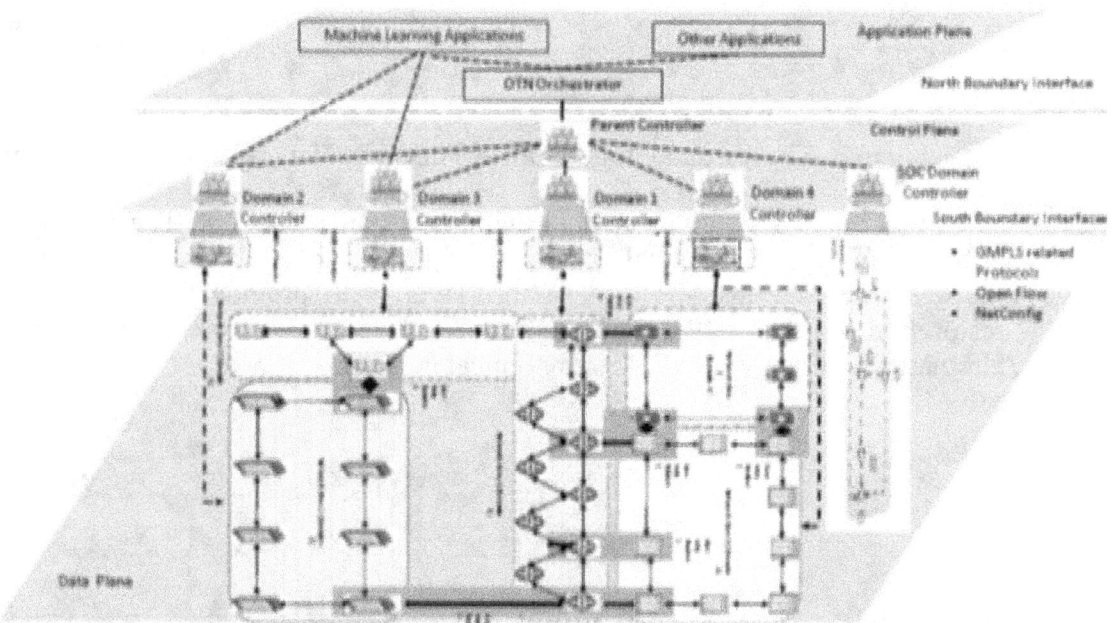

Fig. 6.5. The Prosed Model of the Intelligent Software-Defined Optical Network

By assigning appropriate weight values to every optical link in the different domains, the traffic will be distributed according to the available capacities and the stabilities of the optical links. The SDN orchestrator collects the information is about the performance of the status of every link in the different domains.at the time of the network becomes normal, the SDN parent controller reverts all the restored links again to its main routes after checking its performance [77, 95].

The centralized architecture of the SDN enhances the recovery time in the overall network, where it minimizes the number of node reconfigurations, which needed time for the restoration by bundling the configuration policies to the need domain controller. The parent controller distinguishes between the different paths frailer by using field type where the primary path has n = 0 and secondary paths n > 0; according to the filed type of the failure path, the parent, the controller decides the restoration domain of the network. The principle of the operation is developed by an Open-Flow rule, and based on appropriate protocol, such as the Netconfig/YANG data model as shown in figure 6.6 [95].

5.3.3 The experience of resilience network behaviours

The purpose of the suggested system model of the ML in the long-distance optical network is to observe the changes in the performance of the optical links, by analyzing the actual values of the OSNR in the different optical links with its designed values. The monitoring processes will be done from the records on the change management database (CMDB) of the NMS's of every domain, in case of recognizing any fluctuations in the OSNR values, it will signal these variations and will develop its predictions system according to these data, at the same time the model will advise the optical links for the restoration or for carrying working services, the ML will classify all of the optical links according to the experience of its performance and by using a smart optical performance System (SOPS). The model will cover all the OSNR monitoring duties for the first links in the DWDM network. Also, it will handle end to end optical layer performance in the complete optical network. The training data is used from the change management database (CMDB) of the NMS systems and are grouped according to every Optical port type in the NE as following: Transmit power, receive power, Error rates, Receive End of Live power (EOL), Optical Fiber span attenuation and ONSR

```
augment "/ser:set-service/ser:input" {
    ext:augment-identifier "otn-set-service-input";
    list service-constraint{
        key "index";
        leaf index {type ttypes:constraint-id;}
        uses ttypes:common-constraint;
        leaf service-constraint-type {type ttypes:service-constraint-type;}
        ......
    }
    container source {
        leaf node-ref {type tnode:node-ref;}
        leaf node-connector-ref {type tport:node-connector-ref;}
        leaf access-if-type {type taccessif:access-if-type;}
        list port-constraint{
            key "index";
            leaf index {type ttypes:constraint-id;}
            uses ttypes:common-constraint;
            leaf port-constraint-type {type ttypes:port-constraint-type;}
        }
    }
    container destination {
        ......
    }
}
augment "/ser:service-updated" {
    ext:augment-identifier "otn-service-updated";
    uses otn-service;
}
```

Fig. 6.6. An example of OTN services YANG Data Model

Table 5.1 explains the different types of data which is used as input or output in the SOPS for Every NE [96, 97]. The data set which used to learn the model is a selection from the CMDB of the NMS according to the next parameters:

$$\theta_{att} = 1 \tag{6.6}$$

$$\theta_{rx} = 1 - 1 \tag{6.7}$$

$$OSNR_{Tx} = \frac{P_T}{P_Q F_T} \tag{6.8}$$

$$OSNR_{Rx} = \frac{P_R}{P_Q}\left(\frac{P_N}{P_Q} + \theta_{rx} F_R - 1\right) \tag{6.9}$$

$$\frac{OSNR_{Rx}}{OSNR_{att}} = \frac{\theta_{att}}{\theta_{rx}} = \frac{r}{1-r} \tag{6.10}$$

The evolution of the model will be done according to many examples of the data set which is not used in the learning phase, and the errors will be calculated by using the root mean square errors (RMSE) will be used like the following [28]:

$$RMSE = \sqrt{\frac{1}{N}\sum_{k=1}^{k=N}(\hat{r}_k - r_k)^2} \tag{6.11}$$

The initial phase of formulating the model is the training phase, and the data set in this phase is collected from the CMDB of the NMS for each NE with optical ports with a bit rate of 10 Gb/s or more. There are linear associations between output determinants (OSNR & BER) and the input determinants as following [75, 78]:

$$OSNR = 10\frac{S}{N} \qquad (12)$$

$$10 \log_{10}(BER) = 10.7 - 1.45 \, (OSNR) \qquad (13)$$

$$Y_{1k}(x) = f(\sum_{n=1}^{n=d} \sum_{i=1}^{i=m} a_i x_n^k + w_0) \qquad (14)$$

$$Y_{2k}(x) = f(\sum_{n=1}^{n=d} \sum_{i=1}^{i=m} a_i x_n^k + w_1) \qquad (15)$$

Where: S represents the linear optical signal power, N represents the linear optical noise power, $Y_{1k}(x)$ is the expected output of the OSNR in the port number k, $Y_{2k}(x)$ is the expected output of the OSNR in the port number k, a_i the coefficient value of the hidden layers, x_n^k is the input of n variables for port number k in the network, w_0 is the fixed weight between input variables and output OSNR, AND w_1 is the fixed weight between the input variables and the output BER.

The design model consists of 3 layers, the 1'st one is the input layer which contains n variables and m ports and the 2'nd layer is the hidden layer and forms the association between input and output with p coefficient and the 3'rd layer is the output as shown in figure 6.7 [77].

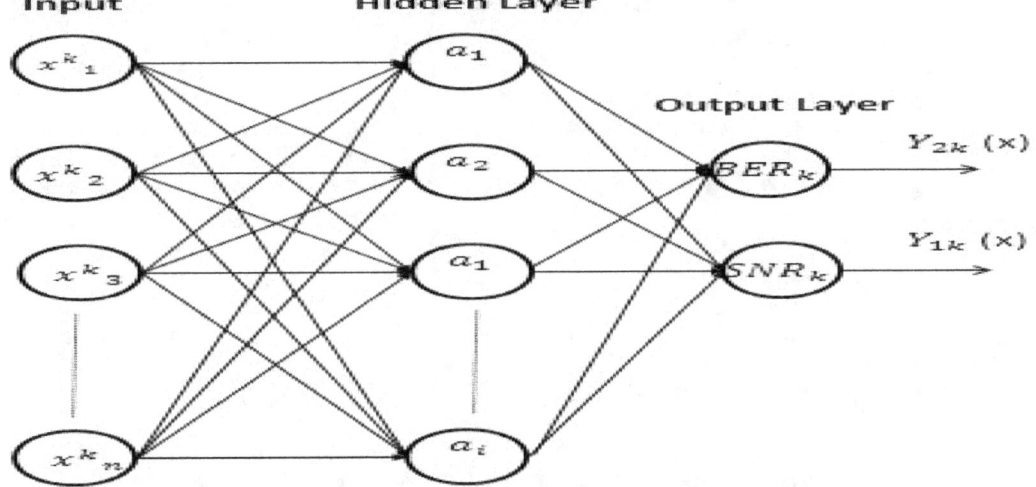

Fig. 6.7. The Proposed Artificial Neural Network for Optical Performance model

5.4 Network scenarios and analysis

There are many scenarios for network failure as following [97, 98]:
- Network failure with one cable cut in the domains of the network To reduce the overload on the parent controller of the SDN the reaction with one failure will be according to the control plane of every domain or according to the controller of the NMS of the domain the only interact from our system is the recommendations from the ML model about the best performance route for the resilience, by sending the best prediction route to the domain controller directly , Adding the control plane uses the LMP, OSPF-TE, and RSVP-TE to realize functions, such as automatic resource

discovery, end-to-end service configuration, and rerouting, on the traditional plane. The signaling module provides the following functions through the RSVP-TE protocol: Sets up or tears down service connections according to user requests, and Synchronizes and restores services according to service status changes. The routing module provides the following functions through the OSPF-TE protocol: Collects and floods the TE link information, and Collects and floods the information about the control link on the control plane, as shown in figure 6.8.

- Computes service trails and control routes. The cross-connection management module provides the following functions: Creates and deletes cross-connections, and Reports information, such as the link status and alarms.

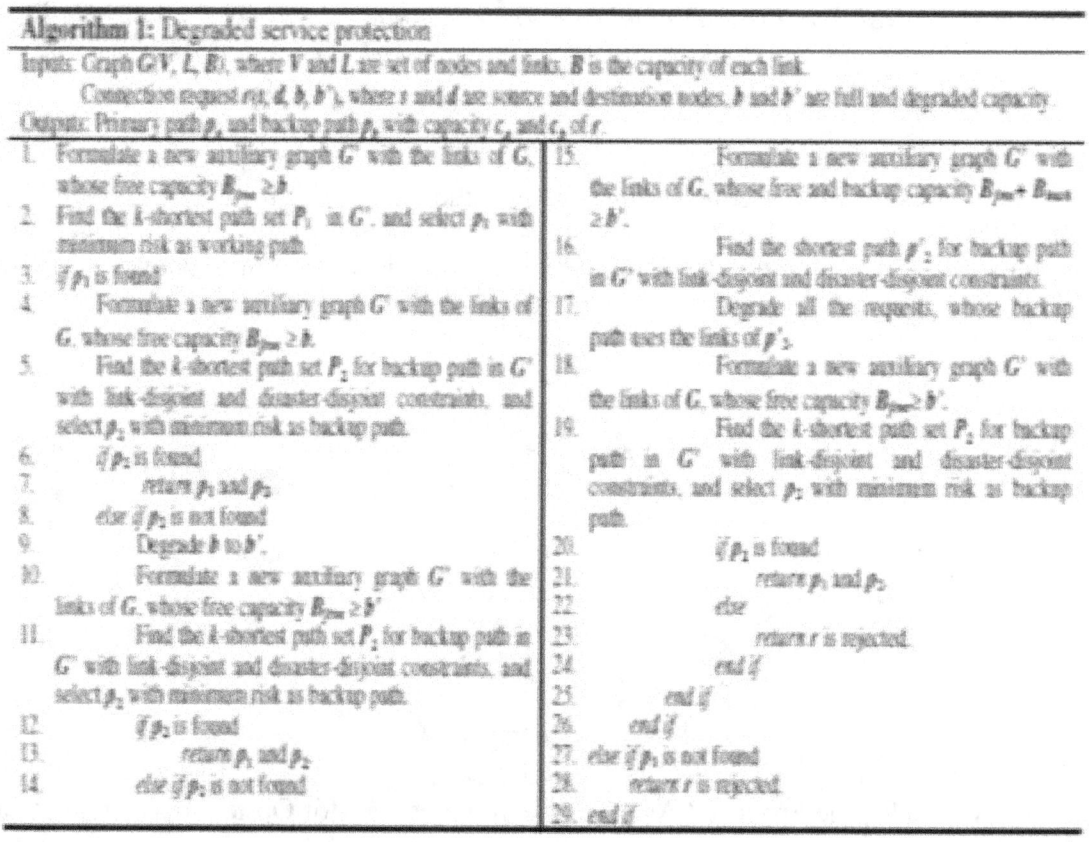

Fig. 6.8. Network failure with more than one cable cut in all domains of the network

The parent controller (PC) of the SDN recognizes failure notification (through some failure detection and scheme), after that PC will try to determine the best possible rerouting trail nearby the failure according to the parameters of the ML output based on the information it has about the current state of the network. After the backup path is estimated, information is sent to all the related domain controllers to reconfigure their switching parts to fit this path. In the multi-layer systems that are recognized, failures are either one of two types: logical, such as an IP service, which is ICC in our model, or optical layer such as VCOTN and SOC in our model. There are thus four possible scenarios, depending on the source of the failure, and the layer that produces the resilience [99, 100]:

- Failure and restoration in the optical-based on the primary routing, as shown in figure 6.9 The connectivity in the optical layer and the resulting connectivity in the logical layer throughout a regular mode of administration. The altered light path is retrieved away from the failure using optical capacity that was stored for this object. The administration is transparent from the logical layer, which stays unchanged.
- Failure and recovery in the logical layer. After failure, the service is rerouted using the residual capacity of the logical layer. The operation is transparent to the optical layer.
- Optical failure repaired in the logical layer, as optical failure recovered in the logical layer. If the optical layer fails to overcome from the optical failure after a specific time-lapse, the logical layer can recover the service on a different logical path, using, for instance, implicit Label Switched Path (LSP) protection in MPLS

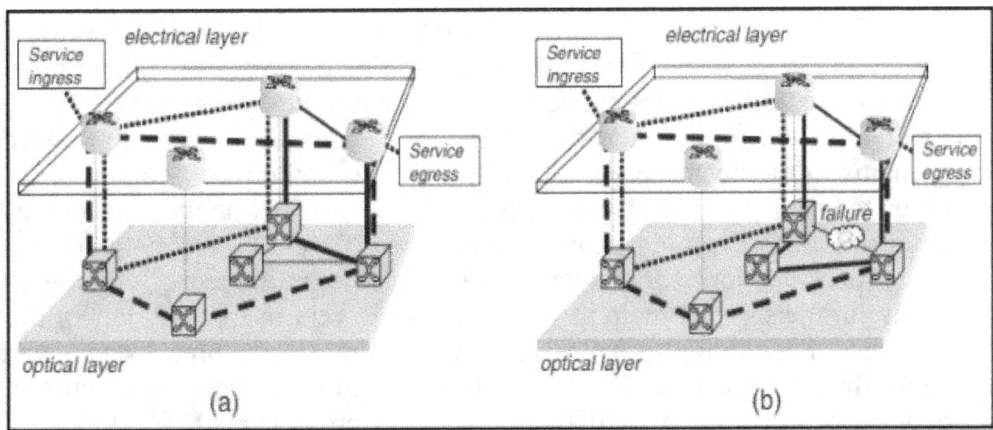

Fig. 6.9. Failure and recovery in the optical-based

Logical failure corrected in the optical layer. Unlike any of the earlier protection schemes, retrieving from a logical failure with an advantage from the optical layer includes reconfiguration with the formulation of new associations in the logical and the optical layer. This type of recovery may be necessary if, after a logical failure, the remaining capacity in the logical layer is insufficient to reroute all the altered services. Further logical capacity can be performed with the provisioning of new lightpaths. The synergy between the restoration's structures used in each layer; the optical layer does not understand a priori the logical connectivity of the client and hence cannot take the ambition to retrieve from a logical failure. Both layers, however, could organize their effort to return [98, 101].

5.6 Conclusion

The future of this work will be a complete simulation by using NS3 and Matlab model to implement the integration between the different functions of SDN, ML, and the OTN. The aim of the ML such the Artificial Neural Network (ANN) is to help the centralized controller of the SDN in the OTN by the experience of the performance of the optical links in the OTN, this enables the centralized controller to assign the right decisions about the optimized routes of the restoration of the service between the different domains and multilayers in the OTN. For the first time, the optical cloud concepts are introduced in the OTN by slicing and virtualizing the various domains with its vendors in the heterogeneous optical network to 3 integrated unified layers, this provides the required resiliencies and the bandwidth on demand between the multilayers and the different domains in the OTN in a more elastic way.

The model is tested using 2 methods , the 1'st method is done by a software simulation by using an SPSS software, and practical experiments to integrate between the different domains of optical network, the centralized controller of the SDN and the ANN to perform the routing and the reliance mechanisms between the multilayers and many domains in the heterogeneous OTN , the 2'nd method was done by performing practical case study on the long distances heterogeneous OTN network in one of the middle east countries about the integrations between the different optical network domains and slicing the optical network to 3 layers to perform the resilience of the services of the multi failure in the same domain through the multilayers in optical network. The results of the new model according to the practical case study in the long-distance heterogamous OTN show that: the dependence on the single vendor is nearly neglected with applying the concept of the clouding and slicing in the heterogeneous OTN, the pay for the end-users bandwidths has become possible and the time to provide the bandwidth on demand has become very short , the meshing between the heterogeneous optical network became available and the resilience for diamond services improved from 25% for double or triple faults to 100% after applying part of our model in the long-distance optical network the resources, the available bandwidth of the optical core network is optimized by more than 25% , the revenue from some OTN domains which have free bandwidths more than 50 % is increased by more than 50%, the switching time enhanced by about 50% and the latency reduced from 27 msec to 742 usec for the selected routes which are optimized from the centralized controller.

Chapter 7
Experimental Results and discussion

7.1 Testing the Security Model

The ability to test the usefulness of the intrusion detection on the optical network and the effectiveness of the proposed security layer will be done as follows:

7.1.1 Test the intrusion detection and response Model

The model is tested by using a software simulator such as the neural designer and the SPSS with 500 records datasets from the practical optical network and contains seven variables, which were collected from the experimental data to train and test the model about any dynamic changes in the BER and the ONSR values. Table 7.1 shows the coefficient of the linear regression model between the detected and the actual outputs and Table 7.2 shows summary of the in-intrusion detection model by using the ANN technique.

Table 7.1: The Coefficients of the Intrusion detection Model

Model		Unstandardized Coefficients		Standardized Coefficients	T	Sig.
		B	Std. Error	Beta		
1	(Constant)	3.62E-09	0		0.291	0.771
	BE	1000	0	1	5090700.1	0
2	(Constant)	1.86E-05	0		5.153	0
	BE	999.977	0.004	1	227553.3	0
	$OSNR_r$	-4.78E-07	0	0	-5.152	0

Table 7.2: The Intrusion Detection Model Summary

Training	Sum of Squares Error	0.344
	Relative Error	0.001
	Stopping Rule Used	One consecutive step(s) with no decrease in error
	Training Time	00:00.4
Testing	Sum of Squares Error	0.429
	Relative Error	0.003

From Table 7.1, the following equation represents the prediction model of the BER.

$$\overline{BER} = 0.357 + 999.977 * ER + -4.776E - 7\ OSNR_r \qquad (7.1)$$

Where: $OSNR_r$ is the optical signal to noise ratio at the transmitter, $\overline{BER}$ is the expected Bit error rates at the receive site, and ER is the error rate at transmitting sites.

Figure 7.1 Shows the Regression between the predicted output values and the actual BER from the intrusion detection model.

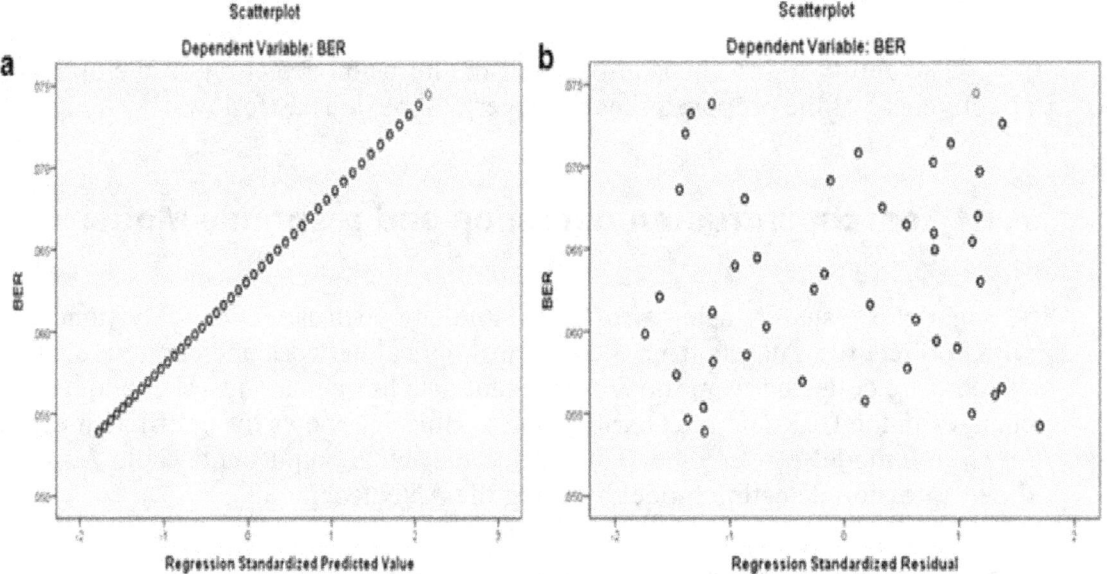

Fig. 7.1. The regression relations between the Prediction and the Actual BER

From equation 7.1 the ANN model will detect any changes in the OSNR values at the receive site, and bit error rate at the transmit site by calculating the expected value of BER. in case of the predicted value of the BER is exceeded the threshold, which is defined by the network administrator, the model will respond by the required action of the intrusion detection.

7.1.2 Testing the proposed security layer in the OTN

By considering the clock rate which is triggered by the machine learning model is equal to the same rate of the OTN frames and estimated by 12.191 usec for OTU_2 also, 1.167 usec for OTU_4 Frame. The OTN system transmits 82027 frames every 1 second for 10 Gb/s, and 598802 frames every 1 second for 100 Gb/s as a result, the dual SRNG will generate 82027 polynomials and initialized secret keys every 1 second in case of 10 Gb/s and 598802 polynomials and initialized secret keys every 1 second in case 100 Gb/s as well. Table 5.3 gives the solutions of equation 4.11 by using Matlab software, which is the primitive polynomial of the SRNG for the polynomial degree with range $10 \leq n \leq 31$ [53].

Table 7. 3: The Number of the Generated Primitive Polynomial for "$10 \leq n \leq 31$"

$\underline{N}$	2^n	ψ (n-1)	P	$\underline{N}$	2^n	ψ (n-1)	P

10	1024	600	60	*21*	2097152	1778112	84672
11	2048	1936	176	*22*	4194304	2640704	120032
12	4096	1728	144	*23*	8388608	8210080	356960
13	8192	8190	630	*24*	16777216	6635520	276480
14	16384	10584	756	*25*	33554432	32400000	1296000
15	32768	27000	1800	*26*	67108864	44717400	1719900
16	65536	32768	2048	*27*	134000000	113000000	4202496
17	131072	131070	7710	*28*	268000000	133000000	4741632
18	262144	139968	7776	*29*	537000000	534000000	18407808
19	524288	524286	27594	*30*	1070000000	535000000	17820000
20	1048576	480000	24000	*31*	2150000000	2150000000	69273666

The generated polynomials increased at a tremendous rate, with every little step increase in the n variable. From table 1 we can find that for the 10 Gb/s bit rate to achieve 82027 polynomials and initialized secret keys in 1 second, n is estimated to be equal to or greater than 21 ($n \geq 21$), and for 100 Gb/s bit rate to achieve 598802 polynomials and initialized secret keys in 1 second, n is estimated to be equal to or greater than 29 ($n \geq 29$). To measure the time required to break the proposed security system for $l_m = l_M / N$, we will consider n = 21 for 10 Gb/s and n = 29 for 100 Gb/s [53]. For 10 Gb/s client signals and n = 21, the length of the ODU_2 equal to 10.037 Gb/s then,

$$N = 10.037 \text{ Gbits/s } / n = 4779 * 10^5 \tag{7.2}$$

For 100 Gb/s client signals and n = 29 the length of the ODU_4 is equal to 104.7944 Gb/s, then

$$N = 104.7944 \text{ Gbits/s } / n = 3742 * 10^6 \tag{7.3}$$

From equations 4.14 and 7.2 the time required to break the security system and to guess the correct key for the encryption process of the 10 Gbit/s client signals in the proposed model by considering n=21, $\tau = 10^{-12}$ sec., and F= 82027 frames/sec will be:

$$T^E \geq F.N.\tau.2^{n-1} = 82027 * 4779 * 10^5 * 10^{-12} * 2^{20} = 475.75 \text{ days} = 1.3 \text{ years} \tag{7.3}$$

From equations 4.14 and 7.3 the time required to break the security system and to guess the correct key for the encryption process of the 100 Gbit/s client signals in the proposed model by considering n = 29, $\tau = 10^{-12}$ sec., and F = 598802 frames/sec will be:

$$T^E \geq F.N.\tau.2^{n-1} = 598802 * 3742 * 10^6 * 10^{-12} * 2^{28}$$
$$= 6961665 \text{ days} = 19073 \text{ years} \tag{7.4}$$

From equations 7.3 and 7.4 we found that to guess one of the keys that will be used in the decryption process of the ODU_k in the proposed security model in the OTN frames, it will take 1.3 years for 10 Gb/s client signals and 19073 years for 100 Gb/s which is not possible in the practical life, and this indicates our proposed system makes it very difficult to break the structure of the OTN frames by any interested hacker in the normal conditions.

Neural Designer Simulator and Statistical Package for the Social Sciences (SPSS) were used to implement the test, and the validity of the intrusion detection by using data set from actual OTN, also Matlab software was used to find a proper polynomial degree from equation 4.11.

7.2 Results of the Power Consumption model

The data set consists of 500 records for seven variables from actual experimental data. The data set is sorted as five inputs (N Number of NE elements, PN Power consumed by NE, PI Power consumed by Infrastructures, PC Power consumed by the cooling system, S weight of the services on the NE) and two outputs (T total consumed power and PUE power active units). Tables 7.4, 7.5, and 7.6 illustrate the correlations and linear regression coefficients between the Input and output Power and the processing model.

Table 7.4: The Correlations between the input variables with outputs

Model		T	N	PN	PI	PC	S
Pearson Correlation	T	1	1	1	0.953	-0.2	1
	N	1	1	.	0.952	-0.2	.
	PN	1	.	1	0.952	-0.2	.
	PI	0.95	0.952	0.952	1	-0.26	0.952
	PC	-0.2	-0.2	-0.2	-0.26	1	-0.2
	S	1	.	.	0.952	-0.2	1
Sig. (1-tailed)	T	.	0	0	0	0	0
	N	0	.	0	0	0	0
	PN	0	0	.	0	0	0
	PI	0	0	0	.	0	0
	PC	0	0	0	0	.	0
	S	0	0	0	0	0	.
N	T	498	498	498	498	498	498
	N	498	498	498	498	498	498
	PN	498	498	498	498	498	498
	PI	498	498	498	498	498	498
	PC	498	498	498	498	498	498
	S	498	498	498	498	498	498

Table 7.5: The coefficients of the model

Model		Unstandardized Coefficients	
		B	Std. Error
1	(Constant)	2.36E-11	0
	PI	0.102	0

	PC	-8.96E-14	0
	S	0.552	0

Table 7.6: The Case Processing Summary

Parameter	N	Percent
Training	345	99.70%
Testing	1	0.30%
Valid	346	100.00%
Excluded	152	
Total	498	

Equation 7.5 represents the estimated model total power consumption in the optical network, according to the test model of the ANN, which dependents on the power of the infrastructure, power of the cooling system, and consumed power for services in the NE's as the following equation.

$$\hat{T} = 2.355E-11 + .102*PI - 8.956E-14\ PC + .552\ S \qquad (7.5)$$

Figure 7.2 shows the relations between prediction values of the total power consumption with the actual values and the residual values from the test model.

Figure 7.3 and Table 7.7 shows the coefficient and the model summary of the linear regression of the predicted PUE, and in the case and the case summary of the output model. Equation 7.6 indicates the prediction model of PUE in the optical network.

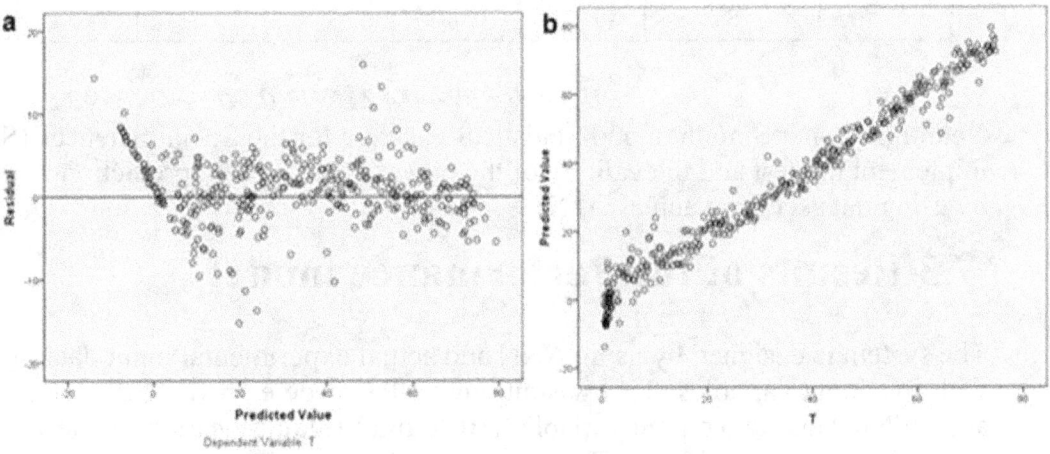

Fig. 7.2. The Regression Relations between the Prediction and the Actual of the total power consumption with the residual

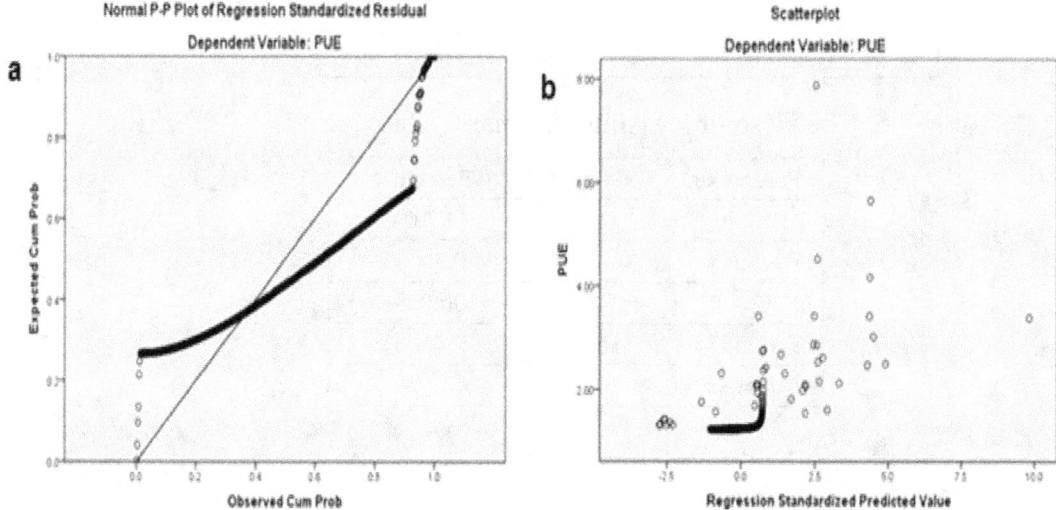

Fig. 7.3. The regression relation between the Prediction Values and the Actual PUE

Table 7.7: The Coefficients and summery of the PUE Test Model

Model		Unstandardized Coefficients		Model Summary			
		B	Std. Error	R	R Square	Adjusted R Square	Std. The error of the Estimate
1	(Constant)	0.357	0.104				
	PI	0.421	0.032				
	PC	0.001	0	0.622	0.386	0.383	0.40324
	S	-0.023	0.001				

$$\widehat{PUE} = 0.357 + 0.421*PI + 0.001\ PC - 0.023\ S \tag{7.6}$$

Neural Designer Simulator and Statistical Package for the Social Sciences (SPSS) were used to implement the test and the validity of the power consumption prediction in the optical network by using data set from actual OTN.

7.3 Results of the Performance model

The system is designed by using ANN and actual experimental input data of 500 records with five input variables (TX transmit power, Rx receive power, ER error rates, OSNRt, and OSNRt) and two output variables (BER and FL fault location), table 7.8 illustrates the correlations and table 7.9 The Coefficients the variables.

Table 7.8: The Correlations between the different variables

		FL	Tx	Rx	SNRt	SNRr	BE
Pearson Correlation	FL	1	.	.	.	.	.
	Tx	.	1	0.816	-0.826	0.746	0.838
	Rx	.	0.816	1	-1	0.411	0.997

	SNRt		-0.826	1	1	-0.412	-0.999
	SNRr		0.746	0.411	-0.412	1	0.412
	BE		0.838	0.997	-0.999	0.412	1

Table 7.9: The Coefficients of the Model

Model		Unstandardized Coefficients		Standardized Coefficients	t	Sig.
		B	Std. Error	Beta		
1	(Constant)	3.62E-09	0		0.291	0.771
	BE	1000	0	1	5090700.061	0
2	(Constant)	1.86E-05	0		5.153	0
	BE	999.977	0.004	1	227553.304	0
	SNRt	4.78E-07	0	0	-5.152	0

Figure 7.31 and equation 7.1 shows the coefficient and the model summary of the linear regression of the performance model.

From the previous results, the estimated values of the BER and the power consumption, which can be calculated by equations 7.1 and 7.6, the same test model can be implemented for the other outputs such fault localization and the configuration management. We can see that the implementation of the total model is applicable by using the ANN technique. Figure 7.4 shows the output from an empirical study in the optical network about the benefits of using the machine learning techniques in performing the operational tasks, and it compares between AS-IS and TO-BE use cases.

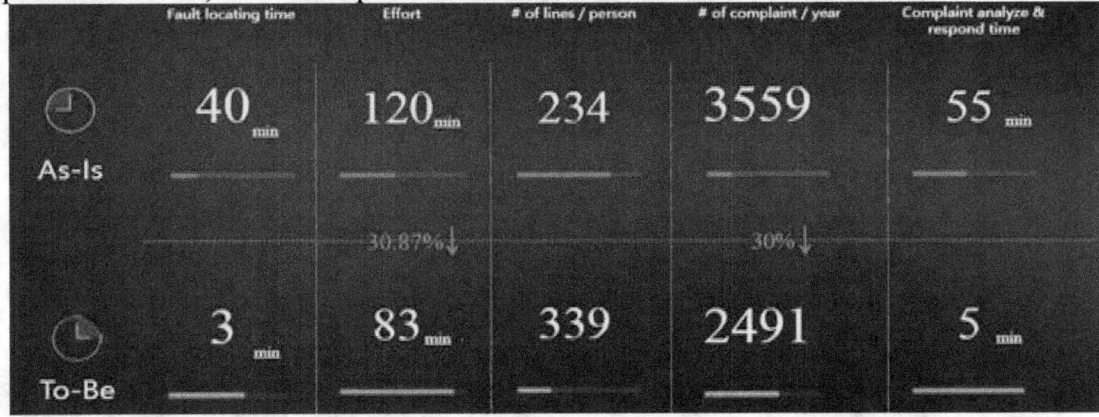

Fig.7.4. An example of the benefits of using AI in the Optical Network

7.4 Results of the Resilience model

A practical case study was done on the real OTN network in one of the middle east countries by applying the integration between 3 different domains with different management systems in the ASON restoration and by simulating only on centralized controller for to supervise the critical services which have strong SLA with the operator of the output results to the significant customers as follows:

- Restoration from more than failure in the different domains in the long-distance optical network for the selected services was improved from 25% for double or triple faults to 100% after applying part of our model in the long-distance optical network.

- The revenue from certain OTN domains which installed in the unpopulated areas and its utilization not more than 25% from the available resource the increased by more than 50% as a result of using the free capacities in theses domains in the restorations of the services from other domains

- The latency from the critical services which will be used in 5G decreased from 27 msec to 742 usec for the selected routes which are optimized from the centralized controller as shown in figure 7.5.

- The switching time of ASON has improved 72 ms to 32 ms by about 50% if the services are sliced in layers as in our model as shown in figure 7.6.

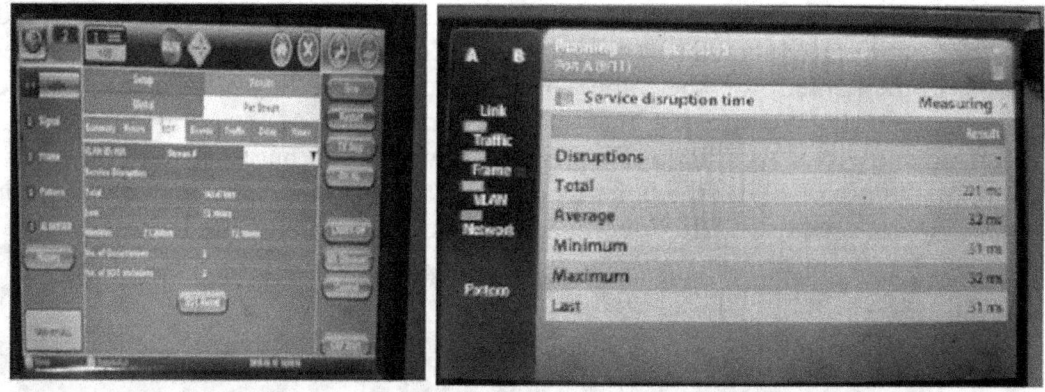

Fig. 7.5. The latency decreased from 27 msec to 742 usec for the selected routes

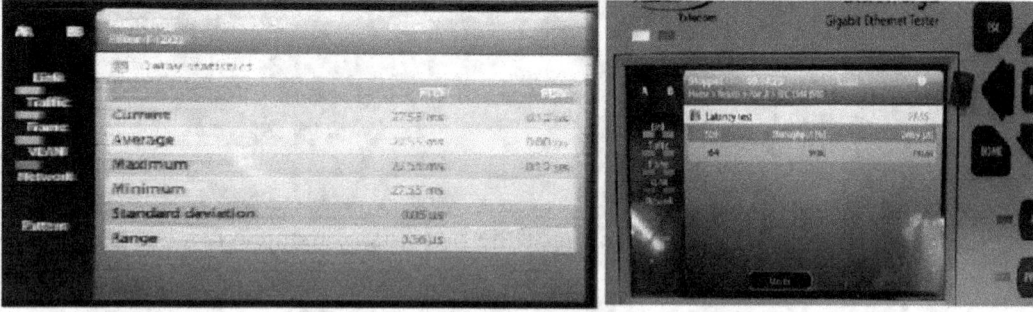

Fig. 7.6. The switching time decreased from 72 ms to 32 ms for the selected routes

Chapter 8
Conclusion

8.1 Conclusion

The book reviewed five significant challenges which faced the Mobile operators while they started to build the traditional 4G network in the backhaul section. These challenges are summarized as the integration with the different network, the capacity of the backhauling links, the energy cost of the macrocells, the security of the 4G network and finally the synchronization of the new 4G network, and we found that there are many solutions were proposed for these challenges which could enhance the performance of the mobile network and open the way to apply the 5G technology by using the existing 4G network. The majority of the proposed solution is concentrated on the virtualization of the network and moves the network from hardware implantation to the software implementations. The virtualization is done by making the decisions of all tasks entire the communication network by a centralized controller for every geographical area. The Software Defined Network (SDN) technology helps in the virtualization of the optical and mobile network by implementing it on the different levels in the communication network. By building the 4G network with the concept of SDN technology will help the operators to overcome many difficulties that they faced during the implementation the network, as all the future solutions will depend on the programmability of the network visualization to perform the integrations between the 4G network with the different IP networks, transform the macrocells to microcells to reduce operational cost of energy, the security algorithms will be implemented in different level of network security not only at access level but also may be at the physical layer of the core network. Merging the artificial intelligence (AI) with the software-defined network (SDN) to manage and control the infrastructure of the mobile network will help in reducing operation cost, monitoring the performance , enhancing the resilience of the 4G services and automatic detection to any intrusions in the network, which have positive impact on the availability and quality of the mobile services at the end-users, also managing the infrastructure of the communication network in smart and centralized way will open the routes to implement the 5G applications with the optimized cost and the required quality, especially with the recent transformations to the IoT ant the electronic governments.

References

1. Yang, H., et al., *Performance evaluation of multi-stratum resources integrated resilience for software defined inter-data center interconnect.* Optics Express, 2015. **23**(10): p. 13384-13398.
2. Sgambelluri, A., et al. *First demonstration of SDN-based segment routing in multi-layer networks.* in *2015 Optical Fiber Communications Conference and Exhibition (OFC).* 2015. IEEE.
3. Skorin-Kapov, N., et al., *Physical-layer security in evolving optical networks.* IEEE Communications Magazine, 2016. **54**(8): p. 110-117.
4. Yan, S., et al. *Field trial of machine-learning-assisted and SDN-based optical network planning with network-scale monitoring database.* in *2017 European Conference on Optical Communication (ECOC).* 2017. IEEE.
5. Alvizu, R., et al., *Comprehensive survey on T-SDN: Software-defined networking for transport networks.* IEEE Communications Surveys & Tutorials, 2017. **19**(4): p. 2232-2283.
6. OIF-VTNS, I., *Virtual Transport Network Services Specification 1.0.* 2017.
7. Brief, O.S., *Openflow-enable transport sdn.* 2014, May.
8. Specification-Version, O.S., *1.4. 0 (Wire Protocol 0x05).* 2013, Open Networking Foundation, Menlo Park, CA, USA.
9. Klaine, P.V., et al., *A survey of machine learning techniques applied to self-organizing cellular networks.* IEEE Communications Surveys & Tutorials, 2017. **19**(4): p. 2392-2431.
10. Ghahramani, Z., *Probabilistic machine learning and artificial intelligence.* Nature, 2015. **521**(7553): p. 452.
11. Reyes, R.R. and T. Bauschert, *Adaptive and state-dependent online resource allocation in dynamic optical networks.* IEEE/OSA Journal of Optical Communications and Networking, 2017. **9**(3): p. B64-B77.
12. Tipmongkolsilp, O., S. Zaghloul, and A. Jukan, *The evolution of cellular backhaul technologies: Current issues and future trends.* IEEE Communications Surveys & Tutorials, 2010. **13**(1): p. 97-113.
13. Bikos, A.N. and N. Sklavos, *LTE/SAE security issues on 4G wireless networks.* IEEE Security & Privacy, 2012. **11**(2): p. 55-62.
14. Jung, H., *Cisco visual networking index: global mobile data traffic forecast update 2010–2015.* 2011, Technical Report, Cisco Systems Inc. 2011. Available online: https://www ....
15. Costa-Requena, J. *SDN integration in LTE mobile backhaul networks.* in *The International Conference on Information Networking 2014 (ICOIN2014).* 2014. IEEE.
16. Sharafat, A.R., et al. *Mpls-te and mpls vpns with openflow.* in *ACM SIGCOMM Computer Communication Review.* 2011. ACM.
17. Aijaz, A., H. Aghvami, and M. Amani, *A survey on mobile data offloading: technical and business perspectives.* IEEE Wireless Communications, 2013. **20**(2): p. 104-112.
18. Sankaran, C., *Data offloading techniques in 3GPP Rel-10 networks: A tutorial.* IEEE Communications Magazine, 2012. **50**(6): p. 46-53.
19. Mobility, I.F., *Seamless Wireless Local Area Network (WLAN) Offload.* Standard 3GPP TS, 2012. **23**.
20. Cui, Q., et al., *A unified protocol stack solution for LTE and WLAN in future mobile converged networks.* IEEE wireless communications, 2014. **21**(6): p. 24-33.
21. Cho, H.-H., et al., *Integration of SDR and SDN for 5G.* IEEE Access, 2014. **2**: p. 1196-1204.

22. Kyung, Y., et al., *Software defined service migration through legacy service integration into 4G networks and future evolutions.* IEEE Communications Magazine, 2015. **53**(9): p. 108-114.
23. Ling, J., et al., *Enhanced capacity and coverage by Wi-Fi LTE integration.* IEEE Communications Magazine, 2015. **53**(3): p. 165-171.
24. Hamada, R.A., H.S. Ali, and M. Abdalla. *Performance evaluation of a novel IMS-based architecture for LTE-WIMAX-WLAN interworking.* in *2014 International Conference on Engineering and Technology (ICET).* 2014. IEEE.
25. Camarillo, G. and M.-A. Garcia-Martin, *The 3G IP multimedia subsystem (IMS): merging the Internet and the cellular worlds.* 2007: John Wiley & Sons.
26. Sánchez-Esguevillas, A., et al., *IMS: The new generation of internet-protocol-based multimedia services.* Proceedings of the IEEE, 2013. **101**(8): p. 1860-1881.
27. Oredope, A., V. Pham, and B. Evans. *Deploying IP Multimedia Subsystem (IMS) services in future mobile networks.* in *2011 National Conference on Communications (NCC).* 2011. IEEE.
28. Liyanage, M., et al. *Leveraging LTE security with SDN and NFV.* in *2015 IEEE 10th International Conference on Industrial and Information Systems (ICIIS).* 2015. IEEE.
29. Cao, J., et al., *A survey on security aspects for LTE and LTE-A networks.* IEEE Communications Surveys & Tutorials, 2013. **16**(1): p. 283-302.
30. Kolias, C., et al., *Openflow-enabled mobile and wireless networks.* white Paper, 2013.
31. Limaye, P. and M. El-Sayed. *Domains of application for backhaul technologies in 3g wireless networks.* in *Networks 2006. 12th International Telecommunications Network Strategy and Planning Symposium.* 2006. IEEE.
32. Korotky, S.K., *Price-points for components of multi-core fiber communication systems in backbone optical networks.* IEEE/OSA Journal of Optical Communications and Networking, 2012. **4**(5): p. 426-435.
33. Douik, A., et al., *Hybrid radio/free-space optical design for next generation backhaul systems.* IEEE Transactions on Communications, 2016. **64**(6): p. 2563-2577.
34. Lometti, A., et al. *Backhauling solutions for LTE networks.* in *2014 16th International Conference on Transparent Optical Networks (ICTON).* 2014. IEEE.
35. Lockie, D. and T. Giaccherini, *Last inch communication system.* 2007, Google Patents.
36. Van Heddeghem, W., et al., *Trends in worldwide ICT electricity consumption from 2007 to 2012.* Computer Communications, 2014. **50**: p. 64-76.
37. Deruyck, M., W. Joseph, and L. Martens, *Power consumption model for macrocell and microcell base stations.* Transactions on Emerging Telecommunications Technologies, 2014. **25**(3): p. 320-333.
38. Humar, I., et al., *Rethinking energy efficiency models of cellular networks with embodied energy.* IEEE network, 2011. **25**(2): p. 40-49.
39. Arshad, M.W., A. Vastberg, and T. Edler. *Energy efficiency gains through traffic offloading and traffic expansion in joint macro pico deployment.* in *2012 IEEE wireless communications and networking conference (WCNC).* 2012. IEEE.
40. De Domenico, A., E.C. Strinati, and A. Capone, *Enabling green cellular networks: A survey and outlook.* Computer Communications, 2014. **37**: p. 5-24.
41. Ha, V.N. and L.B. Le. *Joint coordinated beamforming and admission control for fronthaul constrained cloud-RANs.* in *2014 IEEE Global Communications Conference.* 2014. IEEE.
42. Aweya, J., *Implementing Synchronous Ethernet in telecommunication systems.* IEEE Communications Surveys & Tutorials, 2013. **16**(2): p. 1080-1113.
43. von Zengen, G., et al. *A sub-microsecond clock synchronization protocol for wireless industrial monitoring and control networks.* in *2017 IEEE International Conference on Industrial Technology (ICIT).* 2017. IEEE.
44. Olsson, M., et al., *SAE and the Evolved Packet Core: Driving the mobile broadband revolution.* 2009: Academic Press.
45. Raza, H., *A brief survey of radio access network backhaul evolution: Part II.* IEEE Communications Magazine, 2013. **51**(5): p. 170-177.

46. Raza, H., *A brief survey of radio access network backhaul evolution: part I.* IEEE Communications Magazine, 2011. **49**(6): p. 164-171.
47. Fok, M.P., et al., *Optical layer security in fiber-optic networks.* IEEE Transactions on Information Forensics and Security, 2011. **6**(3): p. 725-736.
48. Furdek, M., et al. *An overview of security challenges in communication networks.* in *2016 8th International Workshop on Resilient Networks Design and Modeling (RNDM)*. 2016. IEEE.
49. Han, M. and Y. Kim. *Unpredictable 16 bits LFSR-based true random number generator.* in *2017 International SoC Design Conference (ISOCC)*. 2017. IEEE.
50. Liu, X.-B., et al., *A study on reconstruction of linear scrambler using dual words of channel encoder.* IEEE Transactions on information forensics and security, 2013. **8**(3): p. 542-552.
51. Dimitriadou, E. and K.E. Zoiros, *All-optical XOR gate using single quantum-dot SOA and optical filter.* Journal of Lightwave Technology, 2013. **31**(23): p. 3813-3821.
52. Mobilen, E., R. Bernardo, and L.R. Monte. *100 Gbit/s optical transport network 40 nm test chip design and prototyping.* in *2017 SBMO/IEEE MTT-S International Microwave and Optoelectronics Conference (IMOC)*. 2017. IEEE.
53. Barlow, G., *A G. 709 Optical Transport Network Tutorial.* Innocor Ltd. Capturado em: http://www. innocor. com/pdf_files/g709_tutorial. pdf, 2003.
54. Loprieno, G. and G. Losio, *Timeslot encryption in an optical transport network.* 2015, Google Patents.
55. Chen, X., et al. *An OSNR calculating method based on network topology for optical network.* in *2017 16th International Conference on Optical Communications and Networks (ICOCN)*. 2017. IEEE.
56. Ji, Y., et al., *Prospects and research issues in multi-dimensional all optical networks.* Science China Information Sciences, 2016. **59**(10): p. 101301.
57. Lohr, J., *Adaptive traffic encryption for optical networks.* 2018, Google Patents.
58. Liyanage, M., et al., *Opportunities and challenges of software-defined mobile networks in network security.* IEEE Security & Privacy, 2016. **14**(4): p. 34-44.
59. Liu, X.-B., et al. *Investigation on scrambler reconstruction with minimum a priori knowledge.* in *2011 IEEE Global Telecommunications Conference-GLOBECOM 2011*. 2011. IEEE.
60. Engelmann, A. and A. Jukan, *Computationally Secure Optical Transmission Systems with Optical Encryption at Line Rate.* arXiv preprint arXiv:1610.01315, 2016.
61. Carter, T., *An introduction to information theory and entropy.* Complex systems summer school, Santa Fe, 2007.
62. Voyant, C., et al., *Machine learning methods for solar radiation forecasting: A review.* Renewable Energy, 2017. **105**: p. 569-582.
63. Suksonghong, K., K. Boonlong, and K.-L. Goh, *Multi-objective genetic algorithms for solving portfolio optimization problems in the electricity market.* International Journal of Electrical Power & Energy Systems, 2014. **58**: p. 150-159.
64. Gao, J., *Machine learning applications for data center optimization.* 2014.
65. Wang, Z., et al., *Failure prediction using machine learning and time series in optical network.* Optics Express, 2017. **25**(16): p. 18553-18565.
66. Panayiotou, T., S.P. Chatzis, and G. Ellinas, *Leveraging statistical machine learning to address failure localization in optical networks.* IEEE/OSA Journal of Optical Communications and Networking, 2018. **10**(3): p. 162-173.
67. Wang, Z., et al., *OSNR and nonlinear noise power estimation for optical fiber communication systems using LSTM based deep learning technique.* Optics express, 2018. **26**(16): p. 21346-21357.
68. Jia, W.-B., et al. *An efficient routing and spectrum assignment algorithm using prediction for elastic optical networks.* in *2016 International Conference on Information System and Artificial Intelligence (ISAI)*. 2016. IEEE.
69. Ayoubi, S., et al., *Machine learning for cognitive network management.* IEEE Communications Magazine, 2018. **56**(1): p. 158-165.
70. Dai, J., et al., *OPTIMUM COOLING OF DATA CENTERS*. 2016: Springer.

71. Rad, P., M. Thoene, and T. Webb, *Best practices for increasing data center energy efficiency.* Dell Power Sol. Mag., 2008: p. 1-5.
72. Ata, R., *Artificial neural networks applications in wind energy systems: a review.* Renewable and Sustainable Energy Reviews, 2015. **49**(534-562).
73. Gosselin, S., et al. *Application of probabilistic modeling and machine learning to the diagnosis of FTTH GPON networks.* in *2017 International Conference on Optical Network Design and Modeling (ONDM).* 2017. IEEE.
74. Tzelepis, D., et al., *Advanced fault location in MTDC networks utilising optically-multiplexed current measurements and machine learning approach.* International Journal of Electrical Power & Energy Systems, 2018. **97**: p. 319-333.
75. Samadi, P., et al. *Quality of transmission prediction with machine learning for dynamic operation of optical WDM networks.* in *2017 European Conference on Optical Communication (ECOC).* 2017. IEEE.
76. Huang, Y., et al., *Dynamic mitigation of EDFA power excursions with machine learning.* Optics express, 2017. **25**(3): p. 2245-2258.
77. Pointurier, Y., *Design of low-margin optical networks.* IEEE/OSA Journal of Optical Communications and Networking, 2017. **9**(1): p. A9-A17.
78. Yan, S., et al. *Field Trial of Machine-Learning-Assisted and SDN-Based Optical Network Management.* in *Optical Fiber Communication Conference.* 2019. Optical Society of America.
79. Taheri, K., et al., *A hybrid artificial bee colony algorithm-artificial neural network for forecasting the blast-produced ground vibration.* Engineering with Computers, 2017. **33**(3): p. 689-700.
80. Yasen, M.Z.Y., R.A.S. Al-Jundi, and N.S.A. Al-Madi. *Optimized ANN-ABC for Thunderstorms Prediction.* in *2017 International Conference on New Trends in Computing Sciences (ICTCS).* 2017. IEEE.
81. Hippert, H.S., C.E. Pedreira, and R.C. Souza, *Neural networks for short-term load forecasting: A review and evaluation.* IEEE Transactions on power systems, 2001. **16**(1): p. 44-55.
82. Jain, A.K., J. Mao, and K.M. Mohiuddin, *Artificial neural networks: A tutorial.* Computer, 1996. **29**(3): p. 31-44.
83. Castoldi, P., et al. *Segment Routing in multi-layer networks.* in *2017 19th International Conference on Transparent Optical Networks (ICTON).* 2017. IEEE.
84. Savas, S.S., et al., *Backup reprovisioning with partial protection for disaster-survivable software-defined optical networks.* Photonic Network Communications, 2016. **31**(2): p. 186-195.
85. Lopez, V., et al. *Towards a transport SDN for carriers networks: An evolutionary perspective.* in *2016 21st European Conference on Networks and Optical Communications (NOC).* 2016. IEEE.
86. Giorgetti, A., et al. *Fast restoration in SDN-based flexible optical networks.* in *Optical Fiber Communication Conference.* 2014. Optical Society of America.
87. Lim, C.-G. and M. Song. *Design and implementation of optical transport network models with path computation.* in *2018 20th International Conference on Advanced Communication Technology (ICACT).* 2018. IEEE.
88. Fontes, H., et al., *Improving ns-3 emulation performance for fast prototyping of routing and SDN protocols: Moving data plane operations to outside of ns-3.* Simulation Modelling Practice and Theory, 2019. **96**: p. 101931.
89. Thyagaturu, A.S., et al., *Software defined optical networks (SDONs): A comprehensive survey.* IEEE Communications Surveys & Tutorials, 2016. **18**(4): p. 2738-2786.
90. Miletić, V., B. Mikac, and M. Džanko. *Modelling optical network components: A network simulator-based approach.* in *2012 IX International Symposium on Telecommunications (BIHTEL).* 2012. IEEE.
91. Beller, D. and H.-J. Jäkel, *Network restoration.* 2012, Google Patents.
92. Jajszczyk, A., *Automatically switched optical networks: benefits and requirements.* IEEE Communications Magazine, 2005. **43**(2): p. S10-S15.

93. Cholda, P., et al., *Quality of resilience as a network reliability characterization tool.* IEEE network, 2009. **23**(2): p. 11-19.
94. Verbrugge, S., et al. *General availability model for multilayer transport networks.* in *DRCN 2005). Proceedings. 5th International Workshop on Design of Reliable Communication Networks, 2005.* 2005. IEEE.
95. Mauthe, A., et al. *Disaster-resilient communication networks: Principles and best practices.* in *2016 8th International Workshop on Resilient Networks Design and Modeling (RNDM).* 2016. IEEE.
96. Moreno, D.F.A., O.J.S. Parra, and D.A.L. Sarmiento. *Heuristic algorithm for flexible optical networks OTN.* in *International Conference on Smart Computing and Communication.* 2017. Springer.
97. Zhao, Y., et al., *SOON: self-optimizing optical networks with machine learning.* Optics express, 2018. **26**(22): p. 28713-28726.
98. Zhang, F., Y. Zuo, and L. Chou. *Research on metro intelligent optical network planning and optimization.* in *2016 15th International Conference on Optical Communications and Networks (ICOCN).* 2016. IEEE.
99. Bouillet, E., et al., *Path routing in mesh optical networks.* 2007.
100. Pióro, M. and D. Medhi, *Routing, flow, and capacity design in communication and computer networks.* 2004: Elsevier.
101. Casellas, R., et al., *Control, management, and orchestration of optical networks: evolution, trends, and challenges.* Journal of Lightwave Technology, 2018. **36**(7): p. 1390-1402.

List of Figures

2.1. The OTN interfaces within the Operator or a Vendor domain	8
2.2. The mapping of the client data to be used in the OTN	8
2.3. The OTN Frame Format	10
2.4. The OPUk frame structure	10
2.5. The ODUk frame structure	10
2.6. The ODUk overhead position in the frame structure	12
2.7. The overhead bytes of the OPUk	12
2.8. The SDN Model Architecture	14
2.9. The SDN Legacy transport network architecture.	15
2.10. The Categories of the ML algorithms	16
2.11. The distinction between probabilistic Gaussian Mixture and k-means	17
2.13. The different applications of the AI in the optical network	19
3.1. The physical and logical layers in the mobile network	23
3.2. The Logical Elements and Control Process	23
3.3. The End-Users Barriers and Mapping to GTP	24
3.4. The Handover Process from MME through S1	24
3.5. The SDN Based Network	25
3.7. The Heterogeneous Network Scenarios	27
3.7. The Hybrid Architecture of SDN and SDR Configurations	27
3.8. The SDN for IP Core Network and Mobile Flatter Architecture	28
3.9. The SDN-CM Interfaces	29
3.10. The Evolution of the Mobile Network and the Key Technologies	29
3.11. The IMS 3-Layer Architecture.	30
3.12. The LTE-WiMAX-WLAN Tight Coupled Interworking Architecture	31
3.13. The 3GPP Security Architecture	33
3.15. The LTE Security Architecture	34
3.16. The SAE Evolved Packet System's (EPS) Architecture	37
3.17. Backhaul Network Technologies	40
3.18. The TDM Backhaul Network of Point-to-Point Leased Lines	41
3.19. The Wavelength division multiplexed passive optical	43

network	
3.20 The total cost of ownership (TCO) in cellular networks	44
3.21. A scalable and cost-effective CRAN Architecture in a challenging dense urban environment	45
3.22. The evolution milestone from conventional 1G to 5G	46
3.23. The Baseband station architecture evolution	46
3.24. The C-RAN LTE Mobile Architecture	47
3.25. The PTP synchronization protocol	51
3.26. The Synchronous Ethernet	52
4.1. The Construction of the OTN Frame	56
4.2. The Structure of the OUT Frame	56
4.3. The Transmission Model of Two Network Elements	57
4.4. The security problem of the Transmission Model	57
4.5. The ANN Model of Intrusion Detection and Response	59
4.6. The proposed Model for the intrusion detection	61
4.7. The Proposed Security Layer in the OTN Frame Structure	62
4.8. The Key Management Entity in the optical network	63
4.9. The Proposed Model of the Security Layer	64
4.9. The Implementation of the security layer in the OTN system with the machine learning model	66
4.10. The Implementation of the security layer in the OTN system with the SDN Intrusion Detection	
4.11. The Structure of the OTN 10/100 Gb/s frame with encrypted ODU	68
5.1. Example of the Transmission Network with Multi-Vendor	71
5.2. The Proposed Artificial Neural Network for power consumption model	74
5.3. The Types of the data in the alarm list of the NMS	76
5.4. The Proposed Artificial Neural Network for fault localization model	77
5.6. The Proposed Artificial Neural Network for the configuration model	80
7. The Proposed Artificial Neural Network for the General Model	82
6.1. Practical Model of the Long-Distance Optical Transport Network	85
6.2. The relation between the ASON-Related Protocols	86
6.3. The Virtual Cloud Optical Transport Network (VCOTN) on the OTN	89
6.4. The Services Optical Cloud (SOC) on the OTN	90
6.5. The Prosed Model of the Intelligent Software-Defined	91

Optical Network	
6.6. An example of OTN services YANG Data Model	92
6.7. The Proposed Artificial Neural Network for Optical Performance model	93
6.8. Network failure with more than one cable cut in all domains of the network	94
6.9. Failure and recovery in the optical-based	95
7.1. The regression relations between the Prediction and the Actual BER	98
7.2. The Regression Relations between the Prediction and the Actual of the total power consumption with the residual	101
7.3. The regression relation between the Prediction Values and the Actual PUE	102
7.4. An example of the benefits of using AI in the Optical Network	103
7.5. The latency decreased from 27 msec to 742 usec for the selected routes	104
7.6. The switching time decreased from 72 ms to 32 ms for the selected routes	104

List of Tables

2.1: The OTU types and capacities	8
2.2: The ODU types and capacities	8
2.3: The OPU types and capacities	9
2.4. The OTUk/ODUk/OPUk frame periods	9
2.4. The OTUk/ODUk/OPUk frame periods	9
2.5: The Payload type code points	11
5.1. The Variables of IOPM.	77
5.2. Total Variables of the universal platform	80
7.1: The Coefficients of the Intrusion detection Model	97
7.2:. The Intrusion Detection Model Summary	97
7.3: The Number of the Generated Primitive Polynomial for "$10 \leq n \leq 31$"	98
7.4: The Correlations between the input variables with outputs	100
7.5: The coefficients of the model	100
7.7: The Coefficients and summery of the PUE Test Model	102
7.8: The Correlations between the different variables	102
7.9: The Coefficients of the Model	103

Acronyms

Optical Transport Network	OTN	edge control plane	ECP
Synchrouns Digital Hierarchy	SDH	virtual network functions	VNF
Software Defined Network	SDN	Session Initiation Protocol	SIP
Network Management System	NMS	Next Generation Networks	NGN
Artificial Neural Network	ANN	., public switched telephone network	PSTN
Artificial Intelligence	AI	Global System for Mobile	GSM
Network Element	NE	Authentication and Key Agreement	AKA
Bit Error Rate	BER	the evolved packet core	EPC
Operational Expenses	OPEX	Radio Resource Control	RRC
International Telecommunication Network	ITU	Evolved Universal Terrestrial Radio Access	EUTRA
Optical Transport Hierarchy	OTH	Key Access Security Management Entries	KASME
Optical Data Unit	ODU	Non-Access Stratum	NAS
Optical Payload Unit	OPU	Network Address Translation	NAT
Optical Transport Unit	OUT	Deep Packet Inspection	DPI
Generalized Multiprotocol Label Switching	GMPLS	theft of service	ToS
Open Flow	ONF	denial-of-service	DoS
Operations Support Systems	OSS	the home location register	HLR
Transport SDN	T-SDN	security algorithms group of experts	SAGE
support-vector machines	SVM	high-speed packet access	HSPA
Markov decision processes	MDP	radio base stations	RBS
Optical signal-to-noise ratio	OSNR	Time Division Multiplexing	TDM
Erbium-Doped Fiber Amplifier	EDFA	capital expense	CAPEX
IP multimedia subsystem	IMS	a line of sight	LOS
evolved Node B	eNodeB	fiber to the home	FTTH
3rd generation partnership project of cellular long term evolution	3GPP LTE	passive optical network	PON

wireless data network	WDN	total cost of ownership	TCO
base stations	BSs	network power consumption	NPC
remote radio heads	RRH	code-division multiple access	CDMA
radio access network	RAN	orthogonal frequency-division multiplexing	OFDM
the internet of things	IOT	primary reference clock	PRC
Mobility Management Entity	MME	enhanced Inter-cell Interference Coordination	EICIC
Serving Gateway	SGW	oven-controlled crystal oscillators	OCXO
Charging Rules Function	PCRF	the network time protocol	NTP
Home Subscriber Server	HSS	precision time protocol	PTP
different Tracking Area ID	TAI	the network time protocol	NTP
General packet Radio Service	GPRS	Synchronous Optical Networks	SONET
user equipment	UE	Time-of-Day	ToD
traffic flow template	TFT	Dense Wave Division Multiplexing	DWDM
packet data network gateway	PDW-GW	Linear Feedback Shift Register	LFSR
General packet Tunneling Protocol	GTP	centralized security controller	CSC
The binding identification	BID	software-defined security	SDS
a home of address	HoA	change management database	CMDB
access network discovery and selection function	ANDF	intrusion provision system	IPS
key performance indexes	KPI	customer edge switches	CES
meantime to repair	MTTR	support vector machine	SVM
Greedy Selection Process	GEP	multi-objective genetic algorithms	MOGA
Open Shortest Path First	OSPF	graph-based correlation	GBC

www.ingramcontent.com/pod-product-compliance
Lightning Source LLC
Chambersburg PA
CBHW060421220526
45465CB00008B/2975